Atlas and Data of Solid-Solution Equilibria of Marine Evaporites

Springer
Berlin
Heidelberg
New York
Barcelona
Budapest
Hong Kong
London
Milan
Paris
Santa Clara
Singapore
Tokyo

E. Usdowski M. Dietzel

Atlas and Data of Solid-Solution Equilibria of Marine Evaporites

With 132 Figures

Springer

Prof. Dr. Eberhard Usdowski
Dr. Martin Dietzel

Universität Göttingen
Geochemisches Institut
Goldschmidtstraße 1
D-37077 Göttingen / Germany

Library of Congress Cataloging-in-Publication Data
Usdowski, E. (Eberhard), 1934 - Atlas and data of solid-solution equilibria of marine evaporites / E. Usdowski, M. Dietzel. p. cm. Includes bibliographical references (p. -) and index.

ISBN 978-3-642-64335-4 ISBN 978-3-642-60284-9 (eBook)
DOI 10.1007/978-3-642-60284-9

1. Evaporites. 2. Salt deposits. 3. Seawater - Composition. 4. Phase diagrams. I. Dietzel, M. (Martin), 1963- . II. Title. QE471.15.E8U83 1998 552.5-dc21 97-33658 CIP

Typesetting: MEDIO, Berlin
SPIN: 10556223 32/3020 - 5 4 3 2 1 0 - Printed on acid-free paper

Preface

Solid-solution equilibria of marine evaporites are important in a wide range of science and technology. However, the data had not yet been summarized in a form that is at the same time comprehensive and permits to understand how the quinary seawater system builds up from its bounding systems. Thus the goal of the present volume is at the same time scientific and educational. The understanding of solid-solution equilibria of the various systems with respect to dissolution, precipitation and transformation of solids, their application to the evolution of brines, and a fast access to data is a necessary requirement for any modelling, especially in Geoscience. Another goal is to show the availability of data. Unfortunately, though solubility data are numereous there are substantial gaps, especially with respect to high temperatures. But also up to about 100 °C data are missing for some of the systems so that they cannot be described entirely. Based on the present volume further work on the solubilities of the minerals of marine evaporites may be promoted.

The data have been viewed and collected over several years by the first author. The second author entered the preparation of the volume when it was realized that besides graphics and tables a fast access to data was required. Although both authors are responsible for the whole volume, responsibility is weighted somewhat differently for the various parts. The first author is more, though not exclusively, responsible for what may be read on paper, whereas the second author is more, though not exclusively, responsible for what may be read on the screen.

The authors should like to acknowledge all those co-workers who have helped at the various stages of the preparation, in particular Mrs. Marion Bell, Dipl. Min. Gerlind Böhme, Mrs. Emilia Cierny, Mrs. Cornelia Friedrich, Dipl. Min. Ariana Hirschfeld, Mrs. Irina Ottenbacher, and Dipl. Min. Holger Schwecke.

We are also grateful to Dr. K. R. Wambach-Sommerfeld, Kaliforschungs-Institut der Kali und Salz GmbH, Germany, for revising the phase diagrams, to Prof. Dr. G. R. Helz, Chemistry Department, University of Maryland, U.S.A., for helpful suggestions and revising the English of Chapter II, and to Dr. I. Stahl, Kaliforschungs-Institut, and Dr. W. Engel, Springer-Verlag, for their co-operation.

Above all we wish to thank Prof. Dr. A. G. Herrmann, formerly at the Geochemisches Institut, Göttingen, and at the Institut für Mineralogie und Mineralische Rohstoffe, Fachgebiet Mineralogie-Geochemie-Salzlagerstätten, Technische Universität Clausthal, Germany, for his permanent input, encouragement, and substantial help in many ways.

Geochemisches Institut
der Georg-August-Universität
Göttingen, December 1996

E. Usdowski
M. Dietzel

Contents

I Introduction

I.1
Purpose of the Volume

The present volume on the phase equilibria of marine evaporites is a new and revised edition of the solubility data of the quinary seawater system Na_2Cl_2-K_2Cl_2-$MgCl_2$-Na_2SO_4-K_2SO_4-$MgSO_4$-H_2O (also denoted NaCl-KCl-$MgCl_2$-Na_2SO_4-H_2O) and its pertinent five quaternary, nine ternary, and six binary systems. The purpose of the volume is to provide an easy and fast access to reliable data and an understanding of the interrelation of the various systems. It has been organized into the sections

- Derivation of Phase Diagrams from Solubility Data,
- Data and Diagrams for Solid-Solution Equilibria, and
- The Computer Program SALSYS.

I.2
Use of the Volume

I.2.1
Derivation of Phase Diagrams from Solubility Data

This section addresses itself to the newcomer. The handling of data requires a general view and understanding of phase relationships. Therefore, the data must be transformed and summarized in phase diagrams. Unfortunately, beginning with ternary systems, the phase relationships of multicomponent systems cannot be represented in two-dimensional diagrams so that projections must be used. Therefore, the principles of data transformation and deriving projections are explained. The systems are arranged in the order of increasing components, and for sake of a universal view the general case for a group of systems is treated. In addition to the derivation of projections the changes of liquid compositions upon the isothermal evaporation of water and the precipitation of solids are explained. Finally, as numerical results are required in any quantitative procedure, examples for calculations are given with respect to a specific system.

I.2.2
Data and Diagrams for Solid-Solution Equilibria

This section is preceded by a table summarizing the mineral names and chemical compositions of the solids which occur in the quinary seawater system, and by a table showing the concentrations of the major components of modern seawater. This is followed by a presentation of the groups of the individual systems in the order of increasing components. For any of the systems three-dimensional views of phase diagrams are shown. In order to elucidate phase relationships the diagrams are not drawn to scale. The true extents of the stability fields are shown by a number of polythermal and isothermal projections. Most of the diagrams cannot be found in the previous literature. They are presented here for the first time. The diagrams are juxtaposed by tables which summarize the liquid compositions at the invariant points of a system. Finally, those polythermal and isothermal phase relationships for which reliable liquid compositions are available , and which may be obtained by the attached computer program, have been marked and summarized.

I.2.3
The Computer Program SALSYS

The program calculates the solubilities for the equilibria of the quinary seawater system and the pertinent five quaternary, nine ternary, and six binary systems. The main menu displays the principal groups of the systems. After choosing one of the groups, a specific system may be selected and other menus are displayed subsequently. For the polytherms the temperature is entered, and both the concentrations of all components for any equilibrium solution and the pertinent associations of solids are displayed. With respect to the isotherms the concentration of one of the components is entered, and the concentrations of the pertinent components are displayed. The concentrations are given in terms of "moles of anhydrous salts dissolved in 10^3 mol of water" (mol/10^3 mol H_2O).

Moreover, a list of the liquid compositions at the invariant points or at the isothermal invariant points may be obtained. These data may also be found in Chapter III. However, beginning with quaternary systems SALSYS displays the data in terms of "moles salt/10^3 mol H_2O" whereas in Chapter III they are given in terms of ion fractions. Finally, the option "Conversion" permits the calculation of combinations of concentrations for the quaternary systems and the quinary system.

More information and instructions for the installation of the program may be obtained from the file README.TXT. The program is started with the command SALSYS. Although it is self-explaining some examples for the various types of systems are given below.

I.3 Handling of the Volume

I.3.1 General Procedure

The newcomer should first study Chapter II. The expert may go directly to Chapter III, choosing the system of interest. There, polythermal and isothermal diagrams are given which depict the phase relationships. Besides this tables are given which summarize the compositions of the solutions at the various invariant points, as well as a list of phase assemblages for which the liquid compositions can be calculated. The attached computer program offers multiple choices.

I.3.2 Examples

I.3.2.1 Binary Systems

System 2.1 Na_2Cl_2-H_2O

The temperatures and liquid compositions at the invariant points and the lines for which solubilities may be obtained from SALSYS are summarized on page 86. Turning to the program, following the path "Binary → Na_2Cl_2", and entering 25 °C, the display shows that halite is in equilibrium with a saturated solution containing 55.5 mol Na_2Cl_2 dissolved in 10^3 mol H_2O (55.5 {Na_2Cl_2}). Entering the temperature at the invariant point c: 0.2 °C equilibrium between halite and hydrohalite is indicated by identical liquid compositions. If a temperature of –10 °C is entered it is displayed that ice is stable with a solution containing 24.7 {Na_2Cl_2}, and that hydrohalite is stable with a solution containing 51.0 {Na_2Cl_2}. Entering –21.2 °C yields the concentration at the invariant point b: ice + hydrohalite.

Selecting the path "Binary → Invariant points" a list is displayed which shows the temperatures and liquid compositions at the invariant points for all binary systems. The abbreviations for the solids are summarized in the file README.TXT.

System 2.3 $MgCl_2$-H_2O

The liquid compositions and temperatures at the invariant points and the lines for which solubilities may be calculated are summarized on page 90. Due to the lack of data, the compositions along the line c–d, at which $MgCl_2 \cdot 8H_2O$ is stable, cannot be obtained. Following the path "Binary → $MgCl_2$" of SALSYS and entering 120 °C yields the solubility of $MgCl_2 \cdot 4H_2O$. Entering the temperature at the invariant point e: 116.7 °C it is displayed that $MgCl_2 \cdot 4H_2O$ and bischofite have the same solubility and are thus stable with each other. If 50 °C is entered the concentration of a saturated solution is given which is in equilibrium with bischofite. Entering –1 °C it is displayed that ice is in equilibrium with a solution of 2.8 {$MgCl_2$}, and that bischofite is stable with a solution containing 99.1 {$MgCl_2$}. If the temperature at the invariant point d: –3.4 °C is entered it is indicated that ice is stable with a solution containing 9.0 {$MgCl_2$}. It is also shown that bischofite is in equilibrium with a solution of 99.0 {$MgCl_2$}. Actually, at –3.4 °C bischofite is stable also with $MgCl_2 \cdot 8H_2O$. However, as no data exist for the line c–d the name of this solid is not displayed. The same is true for –16.8 °C where ice is in equilibrium with a solution containing 36.0 {$MgCl_2$}, and the assemblage $MgCl_2 \cdot 12H_2O$ + $MgCl_2 \cdot 8H_2O$ is stable with a solution of 87.5 {$MgCl_2$}. It should be noted that this applies to all subsequent cases in which compositions at a line, evolving from an invariant point, cannot be calculated.

I.3.2.2
Ternary Systems

Beginning with the ternary systems the phase relations are presented in a standard arrangement. At first three-dimensional, polythermal diagrams and their projections are shown. This is followed by isothermal sections or their projections. The invariant points are denoted by letters, whereas the points pertinent to the bounding systems are marked by numbers.

System 3.1 Na_2Cl_2-K_2Cl_2-H_2O
On page 98 the planes and lines are indicated at which one or two solids are in equilibrium with saturated solutions. For example, sylvite is stable at the plane 5–6–c–b–a, and halite + sylvite are stable along the line b–c. Further, the solids, temperatures, and liquid compositions at the invariant points of the system and its two bounding systems are summarized. Finally those monovariant lines are listed for which liquid compositions as a function of temperature may be calculated via SALSYS.

The liquid composition at the invariant points in terms of ion fractions are given on page 100. General information on projections of phase relations and relative concentrations of ternary systems may be obtained from Section II.3.1.

On page 102 those isotherms are summarized for which liquid compositions may be calculated. At halite saturation the input is {K_2Cl_2} and the output is {Na_2Cl_2}, whereas at sylvite saturation the input is {Na_2Cl_2} and the output is {K_2Cl_2}.

Turning to SALSYS and following the path "Ternary → Na_2Cl_2-K_2Cl_2 → Polytherms" a temperature may be entered. It appears that at 50 °C halite + sylvite are at equilibrium with a saturated solution containing 44.0 {Na_2Cl_2} + 27.1 {K_2Cl_2}. Entering the temperature at the invariant point b: –2.3 °C it is displayed that the limiting compositons of the lines a–b and c–b are identical, indicating equilibrium between halite + hydrohalite + sylvite. At a temperature of –10 °C hydrohalite + sylvite are stable with a liquid composition of 46.5 {Na_2Cl_2} + 11.1 {K_2Cl_2}. Entering a temperature of –21.2 °C it is shown that ice + hydrohalite are in equilibrium with a solution containing 47.3 {Na_2Cl_2} (invariant point of the bounding system Na_2Cl_2-H_2O). It is displayed also that ice + sylvite and hydrohalite + sylvite are stable with saturated solutions containing 38.5 {Na_2Cl_2} + 10.5 {K_2Cl_2}, and 42.7 {Na_2Cl_2} + 9.7 {K_2Cl_2}, respectively. Entering a temperature of −10.7 °C the liquid composition at the invariant point ice + sylvite of the bounding system K_2Cl_2-H_2O is given, as well as the liquid composition of a ternary solution which is stable with hydrohalite + sylvite. At −22.9 °C the limiting compositions at the lines 2–a, 5–a, and b–a are displayed, which merge into point a: ice + hydrohalite + sylvite.

Following the path "Ternary → Na_2Cl_2-K_2Cl_2 → Isotherms" in SALSYS two choices are offered. Choosing "ha: halite" a list of temperatures, corresponding to those on page 102, is displayed for which liquid compositions at saturation with respect to halite may be calculated. Selecting 50 °C and entering 10.0 {K_2Cl_2} it is shown that the pertinent concentration is 51.7 {Na_2Cl_2}. Thus at 50 °C halite may be stable with a saturated solution containing 10.0 {K_2Cl_2} + 51.7 {Na_2Cl_2}. Entering at the same temperature 5.0 {K_2Cl_2} it is displayed that halite is stable with a solution of 5.0 {K_2Cl_2} + 54.1 {Na_2Cl_2}. Analogous results are obtained with respect to sylvite saturation if the path "Isotherms → sy" is chosen. At 50 °C sylvite may be stable with a solution containing 10.0 {Na_2Cl_2} + 45.3 {K_2Cl_2}.

Following the path "Ternary → Na_2Cl_2-K_2Cl_2 → Invariant Points" a table listing the invariant points of the system and the two bounding systems is displayed.

System 3.4 Na_2SO_4-K_2SO_4-H_2O

Beginning with page 114 the first two pages show the three-dimensional, polythermal phase diagram, the solids which are stable with ternary solutions at the bivariant planes, along the monovariant lines, and at the invariant points. Further, the liquid compositions and temperatures at the invariant points and the lines for which liquid compositions can be calculated by SALSYS as a function of temperature are summarized. The next two pages show a projection of the phase relations and the compositions at the invariant points in terms of ion fractions. On the following pages isothermal diagrams are shown, as well as liquid compositions which may be obtained via SALSYS for a given solid, temperature and one of the two concentrations. With respect to mirabilite and thenardite the input is {K_2SO_4}, and the output is {Na_2SO_4}. For arcanite and glaserite the input and output are reversed.

All calculations with SALSYS are analogous to those described for the system Na_2Cl_2-K_2Cl_2-H_2O.

I.3.2.3
Quaternary Systems

System 4.1 Na_2Cl_2-K_2Cl_2-$MgCl_2$-H_2O

General information on quaternary systems with a common ion may be obtained from Section II.4.1. On page 160 the rooms, planes, and lines are indicated where one, two or three solids are in equilibrium with saturated quaternary solutions.

The juxtaposed diagram is presented in terms of ion fractions. Quaternary and ternary points are denoted by letters und numbers, respectively. Within the space 7–8–12–11–c–d–b–a carnallite is stable, at the plane a–b–d–c carnallite + halite are stable, and the line c–d indicates stability of carnallite + halite + sylvite. Invariant points do not exist in the temperature range under consideration. SALSYS calculates the liquid composition for the phase assemblages bischofite + carnallite + halite, and carnallite + halite + sylvite, as a function of temperature.

On the following two pages a projection of the system at Na_2Cl_2 saturation is shown, and two liquid compositions are given in terms of ion fractions. The derivation of expressions for the F-, C_P-, and C_T-values is explained in Section II.4.1.

On page 164 the temperatures of isotherms are summarized for which liquid compositions at saturation with respect to halite + sylvite may be calculated. The pertinent diagrams show the compositions of solutions in terms of $\{Na_2Cl_2\}$-$\{MgCl_2\}$ and $\{K_2Cl_2\}$-$\{MgCl_2\}$ plots.

Following the path "Quaternary → Na_2Cl_2-K_2Cl_2-$MgCl_2$ → Polytherms" of SALSYS and entering 60 °C it is shown that halite + sylvite + carnallite are stable with a solution comprising 4.3 $\{Na_2Cl_2\}$ + 8.8 $\{K_2Cl_2\}$ + 78.6 $\{MgCl_2\}$, and that halite + bischofite + carnallite are in equilibrium with a solution containing 0.9 $\{Na_2Cl_2\}$ + 0.5 $\{K_2Cl_2\}$ + 115.5 $\{MgCl_2\}$. Choosing the path "Quaternary → Na_2Cl_2-K_2Cl_2-$MgCl_2$ → Isotherms", entering 50 °C and 10.0 $\{MgCl_2\}$ the pertinent concentrations 36.0 $\{Na_2Cl_2\}$ and 23.7 $\{K_2Cl_2\}$ are obtained.

System 4.2 Na_2SO_4-K_2SO_4-$MgSO_4$-H_2O

The compositions of the solids in terms of ion fractions are summarized on page 166. On the following pages the rooms, planes, lines, and points are indicated where one, two, three or four solids are in equilibrium with saturated quaternary solutions. The quaternary points are denoted by letters, whereas the points pertinent to the bounding systems are marked by numbers. On page 168 it may be seen that bloedite is stable within the boundaries given by the points 15–16–17–22–21–20–g–l–q–p–m–j–h–e–f. On the plane f–g–l–q–p–i bloedite is in equilibrium with glaserite (p. 170), and along the line p–q bloedite is stable with glaserite and loeweite (p. 172). The invariant points of the bounding systems are summarized on page 174, and on page 176 the temperatures and compositions at the quaternary points are given in terms of ion fractions. The concentrations in terms of "moles salt dissolved in 10^3 mol H_2O" may be obtained from SALSYS choosing the path "Quaternary → Na_2SO_4-K_2SO_4-$MgSO_4$ → Invariant Points". Unfortunately, there are not enough reliable data, so that liquid compositions at the univariant lines cannot be calculated as a function of temperature.

Unlike the system Na_2Cl_2-K_2Cl_2-$MgCl_2$-H_2O the isotherms are presented in terms of ion fractions (pp. 178 – 184). The compositions at the isothermal invariant points in terms of "moles salt dissolved in 10^3 mol H_2O" may be obtained from SALSYS choosing the path "Quaternary → Na_2SO_4-K_2SO_4-$MgSO_4$ → Isotherm Invariant Points". The univariant lines for which compositions may be calculated are indicated on the pages juxtaposing the diagrams. Following the path "Quaternary → Na_2SO_4-K_2SO_4-$MgSO_4$ → Isotherms" the pairs of solids are displayed for which liquid compositions may be obtained. Selecting for example, arcanite + glaserite a list of temperatures is shown. Entering 55 °C and subsequently 20.0 {$MgSO_4$} the pertinent concentrations 6.9 {Na_2SO_4} and 18.7 {K_2SO_4} are displayed.

System 4.3 Na_2Cl_2-K_2Cl_2-Na_2SO_4-K_2SO_4-H_2O
General information on reciprocal salt pairs and the projection of phase relations onto a square may be obtained in Section II.4.2. The description of the above system begins with a table and a diagram showing the compositions of the solids in terms of ion fractions (p. 186). The rooms, planes, and lines where one, two, or three solids are in equilibrium with saturated quaternary solutions are indicated on page 188. It may be seen that arcanite is stable within the space 7–3–19–18–8–10–e–f–27–26. At the plane 15–16–b–21 halite is in equilibrium with hydrohalite, and the line a–f indicates that ice is stable with mirabilite and sylvite. The phase assemblages, temperatures, and liquid compositions at the invariant points are summarized on pages 190 and 192.

Turning to SALSYS, following the path "Quaternary → Na_2Cl_2-K_2Cl_2-Na_2SO_4-K_2SO_4 → Polytherms", and entering a temperature, the concentrations of saturated solutions for the phase assemblages at the monovariant lines given on page 190 will be displayed. For example, entering 30 °C shows that the assemblage glaserite + halite + thenardite is in equilibrium with a solution containing 44.3 {Na_2Cl_2} + 10.7 {K_2Cl_2} + 14.3 {Na_2SO_4}. Glaserite + halite + sylvite is stable with a solution comprised of 43.5 {Na_2Cl_2} + 21.6 {K_2Cl_2} + 4.4 {Na_2SO_4}, and arcanite + glaserite + sylvite are stable in a solution with 14.1 {Na_2Cl_2} + 36.1 {K_2Cl_2} + 2.8 {Na_2SO_4}. Entering the temperature at the invariant point c: 16.3 °C the liquid compositions for the assemblages 1: glaserite + halite + sylvite and 2: arcanite + glaserite + sylvite are displayed as well as the limiting compositions of the lines c–12: glaserite + halite + thenardite and c–d: glaserite + halite + mirabilite which merge into c: glaserite + halite + thenardite + mirabilite. The liquid compositions at the invariant points (p. 192) may be obtained in terms of "moles salt dissolved in 10^3 mol H_2O" by choosing the path "Quaternary → Na_2Cl_2-K_2Cl_2-Na_2SO_4-K_2SO_4 → Invariant Points".

The following pairs of pages show polythermal projections of the system at saturation with respect to halite (p. 194), arcanite (p. 196), sylvite (p. 198), and the sulfates of sodium, mirabilite and thenardite (p. 200). Any point within the diagram indicates the relative composition of a quaternary solution which is stable with two, three, or four solids. The derivation of the expressions for relative compositions is explained in Section II.4.2.

On the following pages isothermal sections of the system are shown in terms of ion fractions (opposite p. 202 to 212). The liquid compositions at the monovariant lines may be obtained from SALSYS by following the path "Quaternary → Na_2Cl_2-K_2Cl_2-Na_2SO_4-K_2SO_4 → Isotherms" and choosing a pair of solids. Selecting glaserite + sylvite it may be seen that data may be obtained for a variety of temperatures. Entering 50 °C and subsequently 35.0 {K_2Cl_2} the pertinent concentrations 26.7 {Na_2Cl_2} and 3.4 {Na_2SO_4} are displayed. The concentrations in terms of "mol/10^3 mol H_2O" at the isothermal invariant points can be obtained following "Quaternary → Na_2Cl_2-K_2Cl_2-Na_2SO_4-K_2SO_4 → Isotherm Invariant Points" and choosing a temperature.

For convenience a set of concentrations of a quaternary solution may be converted into a different set following the path "Conversion → Reciprocal → Data" and entering concentrations. The method of calculation is described in Section II.4.2. For example, defining {K_2Cl_2}=AX, {Na_2Cl_2}=BX, {K_2SO_4}=AY=0, {Na_2SO_4}=BY, and entering the above data yields the combinations

AX	BX	AY	BY
31.6	30.1	3.4	0
35.0	26.7	0	3.4
61.7	0	−26.7	30.1
0	61.7	35.0	−31.6

Data for the systems Na_2Cl_2-$MgCl_2$-Na_2SO_4-$MgSO_4$-H_2O and K_2Cl_2-$MgCl_2$-K_2SO_4-$MgSO_4$-H_2O may be obtained in the same way. However, with respect to the last system there are not enough reliable data for equilibria between three solids and solutions so that concentrations as a function of temperature cannot be calculated.

I.3.2.4
The Quinary System

Unlike the other systems isotherms are presented first for the quinary system. The derivation of expressions for projections of phase relations is explained in Section II.5. The stability fields with respect to the three-dimensional 25 °C-isotherm are illustrated opposite page 278. At any of the planes, lines, or points halite is in equilibrium with one, two, or three solids and saturated, quinary solutions. The capital letters are notations according to D'Ans (1933). On page 280 the compositions at the isothermal invariant points are listed in terms of ion fractions. The fraction of chloride and the water values for the above points along the lines F to Z and G to Z are given as a function of the concentration of Mg^{2+} on pages 282 to 286.

The compositions of the solids in terms of F-values are shown on page 288, and the following pages present isothermal sections at saturation with respect to halite for 15°, 25°, 35°, 55°, 69 °C, and 78 °C. The concentrations in terms of "moles salt dissolved in 10^3 mol H_2O" may be displayed by SALSYS following the path "Quinary → Isotherm Invariant Points" and entering a temperature. For convenience, a set of concentrations may be converted into another set by following the path "Conversion → Quinary system → Data" and entering concentrations. Subsequently, different sets of concentrations are displayed. The conversion of concentrations of a quinary solution is explained in Section II.5.

On page 302 phase relations corresponding to those of the line F–M–N–P–Q–R–Z at 25 °C and to analogous lines at other temperatures are given. Temperatures and compositions of solutions at the invariant points in terms of ion fractions are summarized on page 304. Further below on the same page the lines are listed for which liquid compositions can be calculated by SALSYS.

Phase relations corresponding to those of the line G–S–T–U–V–W–X–Y–R–Z at 25 °C and to analogous lines at other temperatures are given on pages 306 and 308. The compositions and temperatures at the invariant points and lines for which concentrations may be calculated are summarized on pages 310 and 312.

Selecting the path "Quinary → Polytherms" and entering a temperature the compositions of solutions which are in equilibrium with three solids and halite may be calculated. For example, entering 40 °C yields the liquid compositions for the assemblages

1:	glaserite	+ leonite	+ sylvite	+ halite
2:	kainite	+ leonite	+ sylvite	+ halite
3:	carnallite	+ kainite	+ sylvite	+ halite
4:	carnallite	+ kainite	+ kieserite	+ halite
5:	bischofite	+ carnallite	+ kieserite	+ halite
6:	bloedite	+ glaserite	+ thenardite	+ halite
7:	bloedite	+ glaserite	+ leonite	+ halite
8:	kainite	+ kieserite	+ loeweite	+ halite

The concentrations at the invariant points in terms of "moles salt dissolved in 10^3 mol H_2O" may be obtained from SALSYS following the path "Quinary → Invariant Points".

I.4
Synopsis of Marine Evaporites and Solubility Data

I.4.1
Salt Deposits

Evaporites constitute a huge resource of raw materials for a large number of chemical compounds. Initially, interest focussed on sodium chloride which in former time was, and is still today, widely in demand as a preservative for many means of subsistence (e. g. pickled meat, fish, and cabbage). With the beginning of the industrial age widely used chemical compounds such as chlorine, sodium carbonate, -hydroxide, and -sulfate were produced from sodium chloride, so that the exploitation of salt deposits started on a large scale. Moreover, salts other than sodium chloride became more and more interesting, in particular potassium-bearing salts which are used in a number of fertilizers. Recently, artificial caverns in salt deposits have been used for the storage of crude oil and natural gas, and former salt mines have become interesting for the storage of radioactive and chemical wastes, in particular in central Europe (Herrmann and Knipping 1993).

Thus the research on evaporites combines the interests of a variety of different fields: geology, mineralogy, geochemistry, chemistry, agriculture, mining and chemical engineering, economy, and ecology.

The majority of the world's salt deposits is of marine origin. The cations of seawater have been accumulated via continental weathering and water–rock interaction at ocean ridges, whereas the anions and the water have been derived from volcanic gases, essentially (Wedepohl 1995). It is generally agreed that the proportion of the major ions has been virtually constant since Paleozoic time. On the other hand, the concentration of seawater was different, most probably, in different periods of the earth's history, due to climatic variations. Table 3.2 summarizes the major components of modern seawater (Culkin 1965).

Upon the evaporation of seawater various solids will be precipitated subsequently in the order of increasing solubility, generally speaking. Beginning with the crystallization of sodium chloride when about 120 g solution out of 1 kg of seawater are left, most of the calcium carbonate and calcium sulfate has been precipitated so that the remaining solution may be considered to constitute a quinary system. The precipitation of salt minerals, other than halite, begins when about 25 g solution out of 1 kg of seawater are left. The geologic history of a salt deposit commences when the primary precipitates are covered by other sediments, e. g. sand and clay. After a period of time the evaporites are altered diagenetically and may be deformed by halokinetic movements. During these processes the primary phase assemblages will be transformed into new assemblages via chemical reactions between the solids and solutions included in the pore space, depending upon the temperature and the composition of the pore fluids. Moreover, meteoric water finding access to a salt deposit may redissolve the solids, producing subsurface brines which may enter the hydrosphere.

I.4.2
Significance of Solubility Data

As the various assemblages of salt minerals and water constitute multi-component systems the primary crystallization of the solids upon evaporation and their subsequent alteration under hydrologically closed or open system conditions follow a complicated path of chemical reactions. Thus any evaluation of the above and other processes must be based on the knowledge of the solid-solution equilibria of the various assemblages. Such data are a necessary requirement for a geochemical and geological modeling of the formation and alteration of salt deposits and the evolution of subsurface brines (Braitsch 1962, 1970; Herrmann et al. 1978, 1979). Besides this the data are also required by a number of other scientists: hydrologists and ecologists who are concerned with the contamination of surface and ground water by solutions running off salt dumps, experimental petrologists studying the interactions between brines and rocks, and industrial engineers who are concerned with the various ways of processing salt rocks, e. g. the separation of potassium chloride from a halite–sylvite rock via dissolution and re-precipitation using temperature gradients. Briefly, solubility data are required by all scientists who are concerned with the formation of salt minerals, the evolution of brines, and their interaction with other materials.

I.4.3 Data Bases

Systematic research on the solid–solution equilibria of the seawater system and its bounding systems started with J. H. van't Hoff near the end of the last century. His famous work was published first in the "Sitzungsberichte der Königlich Preußischen Akademie der Wissenschaften" between 1897 and 1908. Most of the articles were repubished as early as 1912. Since van't Hoff the research on solid–solution equilibria has grown steadily. In 1933 J. D'Ans, who also contributed a considerable amount of important data, published a compilation of salt–solution equilibria concerning the seawater system and its bounding systems, tracing the literature back to 1793. This was succeeded in 1942 by the volume "Kalium", an appendix to "Gmelins Handbuch der Anorganischen Chemie". The volume was updated in 1970. Another important compilation is the voluminous collection of data by Zdanovskij et al. (1973, 1975). Numerous solubility data have entered other handbooks, e. g. D'Ans and Lax (1943), Seidell (1953, 1958, 1965), and Landolt-Börnstein (1962).

Searching for data is laborious, time consuming, and may present difficulties. In particular the newcomer may be at loss in quite a number of cases because the data from different sources are not consistent and even contradictory, different units of concentrations are given, and because it may not be clear to which out of several hydrates the data refer. Moreover, phase relationships may not be depicted sufficiently by the joined diagrams. This does not apply so much to the simple but to the more complicated systems.

In the present compilation the data have been evaluated with respect to their consistency, and if it was possible to their quality. Unfortunately, there are rather little reliable data for higher temperatures. The pertinent literature has been reviewed by Fanghänel and Emons (1992). In the present volume, only data up to about 100 °C are given, in general, and only such data have been considered which refer to solids which are generally accepted as stable phases. The concentrations were plotted as a function of temperature, or with respect to isothermal sections of multicomponent systems the concentration of one of the components was plotted against that of the other components. Regression analyses showed that all polythermal and isothermal relationships are sufficiently described by second-order equations. Higher order or other expressions yielded no better results, in general. Only such data have been considered for which the regression analyses yielded a r-value > 0.98. As a matter of fact in most of the cases the r-values are better than 0.99. The coefficients of the second-order equations for the polythermal and isothermal equilibria have been entered into the computer program SALSYS which provides an easy and fast access to critically evaluated solubility data. The above cited references are given at the end of Chapter I.

I.4.4 Individual Articles

Beginning with the era of van't Hoff, the references for the individual articles have been summarized in the file LIT.DOC of the attached disk. It should be noted that the articles cover a variety of topics and that in many cases emphasis is on subjects other than solubilities, such as vapor pressures, densities of solutions, freezing point depression, luminescence, syntheses of solids, isomorphous replacement, and industrial application. Thus, appointing a publication to a specific system does not imply that this system is completely covered and that it is concerned totally with solubilities. Nevertheless, data may be found. It also should be noted that a good many of the data are not reliable. Usually this is due to the fact that equilibrium is difficult to attain experimentally and that metastable states prevail. However, such data may be useful for further research.

I.4.5 Comments

I.4.5.1 Binary Systems

The solubility data and phase relations of the binary systems have been well established since the turn of the century. The solubilities were plotted as a function of temperature. It appeared that the solubilites can be described sufficiently by one, or if a larger temperature range is considered, by two second order equations. The regression lines yield data in good agreement with the previous literature (e.g. D'Ans, 1933). For temperatures above 100 °C some data may be debatable. With respect to the systems Na_2Cl_2-H_2O and K_2Cl_2-H_2O more high temperature data may be found in Lorimer (1991).

I.4.5.2 Ternary Systems

Tables from the original literature have been reproduced by Zdanovskij et al. (1973). This collection comprises essentially the data used by D'Ans (1933) and data from articles by Russian authors. Polythermal and isothermal regression lines were calculated, excluding erratic and inconsistent data. With respect to the system Na_2Cl_2-$MgCl_2$-H_2O some preference was given to the data by Autenrieth and Braune (1960 a, b). In the system $MgCl_2$-$MgSO_4$-H_2O the stability of f: epsomite + hexahydrite + kieserite has been considered (D'ANS et al., 1960). This assemblage yields point e: bischofite + epsomite + kieserite, which was obtained from the intersections of lines d–e and g–e. The isotherms of the ternary systems can be sufficiently described by a single second-order equation, whereas the polytherms are described by one or two equations, depending upon the temperature range and the individual case.

I.4.5.3
Quaternary Systems

Original data covering the time period between 1881 and 1973 have been reproduced by Zdanovskij et al. (1975). These tables contain not only equilibrium data but also numerous solubilites which refer to metastable states and which may be useful if industrial processes are considered. Those data which may be taken to represent equilibria show that a number of phase relations and solubilites given by D'Ans (1933) are still valid. In other cases new phase relations have been elaborated over the years. On the other hand, quite a number of important data which are required to describe a quaternary system entirely are still not available and are subject to further research.

System 4.1 Na_2Cl_2-K_2Cl_2-$MgCl_2$-H_2O
The data have not changed virtually since D'Ans (1933). In the present volume only phase relations beginning with 0 °C have been considered. However, there are data for the hydrates of sodium chloride and magnesium chloride which are stable below the freezing point.

System 4.2 Na_2SO_4-K_2SO_4-$MgSO_4$-H_2O
The three-dimensional, polythermal diagram is according to Jänecke (1935 a, b, c). However, some modifications had to be made due to inconsistencies of three invariant points with corresponding, incorrect points of two of the bounding systems. With respect to the system 3.5 Na_2SO_4-$MgSO_4$-H_2O Jänecke (1935 a, b, c) proposed for the present point 17: bloedite + hexahydrite + loeweite the phase assemblage bloedite + hexahydrite + kieserite, and for the present point 18: hexahydrite + kieserite + loeweite the assemblage bloedite + kieserite + loeweite. In the system 3.6 K_2SO_4-$MgSO_4$-H_2O the present point 10: hexahydrite + leonite + langbeinite is denoted as hexahydrite + kieserite + leonite, and for the present point 11: hexahydrite + kieserite + langbeinite the assemblage kieserite + leonite + langbeinite was given. Thus the former assemblages bloedite + hexahydrite + kieserite + leonite, kieserite + leonite + langbeinite + loeweite, and bloedite + kieserite + leonite + loeweite of the quaternary system have been replaced by the present points m: bloedite + hexahydrite + leonite + loeweite, n: hexahydrite + leonite + langbeinite + loeweite, and o: hexahydrite + kieserite + langbeinite + loeweite.

The error with respect to the temperatures and liquid compositions at the invariant points given by Jänecke (1935 a, b, c) is difficult to assess. It is assumed that the data are correct within ±1 °C and ±1 mol salt dissolved in 10^3 mol H_2O. The compositions and temperatures for the replaced points are tentative. However, they should not deviate too much from the true values.

The isotherms were obtained from data by Yanat'eva (1949, 25 °C), Abutkova (1972, 25 °C), Bayliss et al. (1947, 35 °C), Yanat'eva and Orlova (1956, 55 °C), and Andronova and Lepeshkov (1958, 75 °C). Data for the 0 °C-, 15 °C-, and 50 °C-isotherms may be found in Yanat'eva and Orlova (1958), Andronova (1958), and Solov'eva and Lyakhovskaya (1969), respectively.

Although there are sufficient data for the isotherms there are not enough points in order to calculate equilibria between three solids and solutions as a function of temperature. It appears that the system is still subject to future research.

System 4.3 Na_2Cl_2-K_2Cl_2-Na_2SO_4-K_2SO_4-H_2O

Points a, c, d, e, and f are given according to D'Ans (1933). The existence of the present point b: halite + hydrohalite + mirabilite + sylvite, not mentioned by D'Ans (1933), is based on the intersections of the lines 16–21 and a-=d (see also Braitsch 1962, 1971). The intersection of the two lines yields a temperature of –2.4 °C. Compositions of this points were derived from second order regressions of points located on the lines a–b (mirabilite + hydrohalite + sylvite) and d–b (halite + mirabilite + sylvite), extrapolating to -2.4 °C.

The data for the isotherms have been taken from Zaslaviskij et al. (1938, -5°, 25°, 75 °C), Yanat'eva and Orlova (1959 a, 0 °C), Solov'eva (1970, 50 °C), and Luk'yanova and Podzorey (1958, 100 °C). For the polythermal regression lines the data by D'Ans (1933) and the data for the isothermal invariant points of the above isotherms have been used.

System 4.4 Na_2Cl_2-$MgCl_2$-Na_2SO_4-$MgSO_4$-H_2O

The majority of the data is from Autenrieth and Braune (1960 a, b). Some of the numerous data of the Russian literature are included. The data which permit the construction of the –5 °C- and 100 °C-isotherms have not been considered. The present point b: bloedite + halite + mirabilite + thenardite is from Pel'sh (1964), and point l: epsomite + halite + hydrohalite + mirabilite is according to Zdanovskij et al. (1975, II-1, p. 469). The temperature and the composition at point j: bloedite + d'ansite + thenardite + vanthoffite have been derived from the intersection of lines j–33 and g–35.

System 4.5 K_2Cl_2-$MgCl_2$-K_2SO_4-$MgSO_4$-H_2O

Most of the data have been taken from D'Ans (1933, 1944) and Autenrieth (1954), including data from the Russian literature. Due to the lack of data the coexistence of epsomite + hexahydrite + kieserite was neglected. The uncertainty of the temperature at the invariant points and of a number of compositions is not better than ±1 °C and ±1 mol salt dissoled in 10^3 mol H_2O. Unfortunately, there are not enough consistent points for three solids and saturated solutions so that the compositions of solutions as a function of temperature cannot be given. Among the data for the isotherms only those which are given here make plausible sense. It appears that the system is subject to further research.

I.4.5.4
The Quinary System

Since D'Ans (1933) a number of articles from essentially Russian authors have appeared, concerning the quinary system at temperatures above 0 °C. Emons et al. (1975 b) have published data for temperatures below the freezing point. The present data have been elaborated over the years by the "Kaliforschungs-Institut" and have been compiled by Usdowski et al. (1997). Within the range from 2° to 108 °C the number of invariant points at saturation with respect to halite has been reduced from originally 32 to 30 by discarding 13 former and introducing 11 new phase assemblages. Moreover, there are revised data for the still valid points as well as new and revised data for six important isothermal sections of the system.

It should be pointed out that the quinary system presents a unique example in so far as any of the solids are stable with halite, exept for arcanite (K_2SO_4). This solid is stable only at undersaturation with respect to halite. Although the phase relations involving arcanite are of comparatively small importance they are required for a number of cases so that research should be continued in this direction. Some data with respect to K_2SO_4 saturation may be found in D'Ans (1933), Yanat'eva and Orlova (1958, 1959 b, 1963), and Solov'eva and Lyakhovskaya (1967, 1968, 1970).

Selected Bibliography

Braitsch, O.: Entstehung und Stoffbestand der Salzlagerstätten. Springer, Berlin Göttingen Heidelberg (1962), 232 pp

Braitsch, O.: Salt deposits. Their origin and composition. Springer, Berlin Heidelberg New York (1971), 297 pp (translation)

Culkin, F.: The major constituents of seawater. In: Riley, J. P., Skirrow. G. (eds.) Chemical oceanography vol 1. Academic Press, London (1965), pp 121–161

D'Ans, J.: Die Lösungsgleichgewichte der Systeme der ozeanischen Salzablagerungen. Verlagsgesellschaft für Ackerbau, Berlin (1933), 254 pp

D'Ans, J., Lax, E.: Taschenbuch für Chemiker und Physiker. Springer, Berlin (1943), 1896 pp

Fanghänel, T., Emons, H. H.: Neue Ergebnisse über die fest-flüssig Gleichgewichte der Systeme der ozeanischen Salze (T > 100 °C). Abh. Sächs. Akad. Wiss. Leipzig, Mat.-nat. Klasse, 57. Akademie Verlag, Berlin (1992), 98 pp

Gmelins Handbuch der anorganischen Chemie. System-Nummer 22, Kalium
Anhangband. Verlag Chemie, Berlin (1942), 220 pp
Ergänzungsband. Verlag Chemie, Weinheim/Bergstr. (1970), 208 pp

Herrmann, A. G., Knipping, B.: Waste disposal and evaporites. Lecture Notes in Earth Sciences 45. Springer, Berlin Heidelberg New York (1993), 193 pp

Herrmann, A. G., Siebrasse, G., Könnecke, K.: Computerprogramme zur Berechnung von Mineral- und Gesteinsumbildungen bei der Einwirkung von Lösungen auf Kali- und Steinsalzlagerstätten (Lösungsmetamorphose). Kali u. Steinsalz 7, 288–299 (1978)

Herrmann, A. G., Siebrasse, G., Könnecke, K.: Computerprogramme zur Berechnung von Thermometamorphose-Prozessen in marinen Salzlagerstätten. Kali u. Steinsalz 9, 389–394 (1979)

van't Hoff, J. H.: Untersuchungen über die Bildungsverhältnisse der ozeanischen Salzablagerungen, insbesondere des Stassfurter Salzlagers (Hrsg.: H. Precht und E. Cohen). Akademische Verlagsgesellschaft, Leipzig (1912), 359 pp

Landolt-Börnstein: Eigenschaften der Materie in ihren Aggregatzuständen, 2. Teil, Lösungsgleichgewichte I (Schäfer, K., Lax, E., eds.). Springer, Berlin Göttingen Heidelberg (1962)

Seidell, A.: Solubilities for inorganic and metal organic compounds. D. van Nostrand Company, New York (1953), 1698 pp
Vol. I (A-Ir) 4th edition (Linke, W. F., ed.). D. van Nostrand Company, New York (1958), 1487 pp
Vol. II (K-Z) 4th edition (Linke, W. F., ed.). Am. Chemical Society, Washington, DC (1965), 1914 pp

Wedepohl, K. H.: Stoffbestand und Entwicklung des Meerwassers. Kali u. Steinsalz 11, 311–315 (1995)

Zdanovskij, A. B., Solov'eva, E. F., Lyakhovskaya, E. I., Shestakov, N. E., Shleymowich, R. E., Abutkova, L. M.: Solubility data for salt-water systems (Pel'sh, A. D., ed.) I. Systems with 3 components. Khimiya, Leningrad (1973), 1070 pp.
II. Systems with 4 and more components. Khimiya, Leningrad (1975), 1063 pp. (in Russian)

II Derivation of Phase Diagrams from Solubility Data

There are two types of quinary salt-water systems (Fig. 2.1). One of them comprises three different cations (A^+, B^+, C^+) and two different anions (X^-, Y^-). These yield the salts AX, BX, CX, AY, BY, and CY. The other one comprises two different cations (A^+, B^+) and three different anions (X^-, Y^-, Z^-) which yield the salts AX, AY, AZ, BX, BY, and BZ. The quinary seawater system Na_2^{2+}-K_2^{2+}-Mg^{2+}-Cl_2^{2-}-SO_4^{2-}-H_2O which yields Na_2Cl_2, K_2Cl_2, $MgCl_2$, Na_2SO_4, K_2SO_4 and $MgSO_4$ belongs to the first type. In the following sections it will be explained how this type originates from its subsystems, in which way the subsystems are related to each other and to the quinary system, and how projections of the various subsystems and the quinary system are obtained. As a matter of fact, all explanations apply likewise to the second type of the quinary systems, in principle.

The explanations are given for a hypothetical quinary system and hypothetical subsystems which exhibit simple phase relationships, so that a comprehension of the quinary seawater system and its bounding systems is facilitated. The systems are treated in the order of increasing numbers of components. For any type of a subsystem a single example is given. This applies to all other systems of the same group, as well.

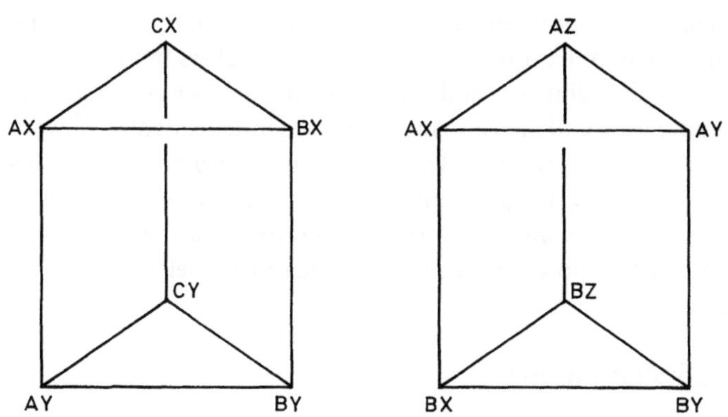

Fig. 2.1 Isotherms of the quinary systems A^+-B^+-C^+-X^--Y^--H_2O and A^+-B^+-X^--Y^--Z^--H_2O

Throughout the text the following nomenclature is used:

AX, BX, etc.:	component or dissolved salt
AX_S, BX_S, etc.:	solid AX, BX
{AX}, {BX}, etc.:	concentration in terms of "moles of salt dissolved in 10^3 mol of water"
C_T:	total concentration (moles of total salts dissolved in 10^3 mol of water)
C_P:	specified concentration
$W_T = 10^5/C_T$:	water value (moles of water in which 100 mol of salt are dissolved)
W_P:	water value with respect to a specified concentration
f:	fraction of ions or dissolved salts with respect to the total concentration
F:	fraction with respect to a specified composition

It should be noted that the concentrations of molecules and ions are interrelated. For example

$$\{AX\} = \{A^+\} = \{X^-\}$$
$$\{AX\} + \{BX\} = \{A^+\} + \{B^+\} = \{X^-\}$$
$$\{AX\} + \{AY\} = \{A^+\} = \{X^-\} + \{Y^-\}$$
$$\{AX\} + \{BY\} = \{A^+\} + \{B^+\} = \{X^-\} + \{Y^-\}$$

II.1 The Phase Rule

It may be worthwhile to apply the phase rule to the one or other system. This rule is expressed by the equation

$$P + F = C + 2 \qquad \text{(II.1)}$$

where P denotes the number of phases (solids, solution, water vapor), F is the number of degrees of freedom or independent variables (temperature, pressure, composition of the solution), and C denotes the number of components. In general the temperature and pressure may be varied independently. However, the present cases are viewed with respect to the presence of water vapor so that the pressure cannot be varied independently. It is given by the water vapor pressure of the solution which depends upon the temperature at a specified concentration. Some examples for applying the phase rule are given below.

II.2 The Binary System AX-H_2O

In Fig. 2.2 point 1 is the melting point of ice (concentration of AX = 0). The melting point is lowered by the addition of AX. The line 1-2 indicates the concentrations of solutions which are in equilibrium with ice and water vapor. As the number of components is C = AX + H_2O = 2 and the number of phases is P = ice + solution + vapor = 3 the degrees of freedom is F = 1. This means that either T or the concentration of AX can be varied independently. Line 1-2 is a monovariant line. At point 2 solid AX becomes stable so that there is equilibrium between ice, AX_s, solution and vapor. As the number of phases is P = 4 the degrees of freedom become F = 0. Point 2 is an invariant point which has a defined concentration and temperature. The monovariant line 2-3-4 indicates compositions of solutions which are in equilibrium with solid AX and vapor (solubility curve of AX_s). Any point above the line 1-2-3-4 represents concentrations of undersaturated solutions. As the number of phases is P = solution + vapor = 2 the degrees of freedom become F = 2. The concentration and the temperature can be varied independently from each other, e. g. the area above the line 1-2-3-4 is divariant (or bivariant).

If water is evaporated under isothermal conditions (constant temperature) from the solution at point 5 the concentration of the solution increases along the line 5-3. At point 3 the solution becomes saturated with respect to AX so that upon further evaporation of water solid AX is precipitated. If a mixture of solution 3 and AX_s is heated, the solid will dissolve and the concentration of the solution will increase, for example to that of point 4. If at this point all AX_s has been consumed and the temperature is increased still the solution enters the divariant field. Upon cooling of any saturated solution solid AX is precipitated. The concentration of the solution changes as given by the line 2-3-4.

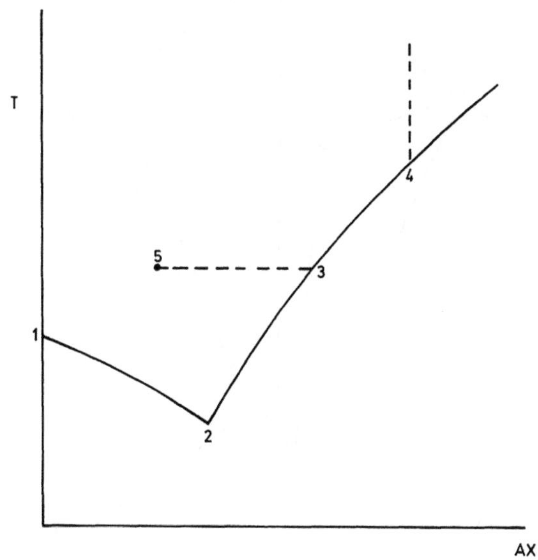

Fig. 2.2 System AX-H_2O. Concentration of component AX as a function of temperature. 1-2: ice + solution. 2-3-4: AX_s + solution (solubility curve of AX_s). 2: ice + AX_s + solution. Area above 1-2-3-4: unsaturated solutions

In the above example AX_s has a positive solubility coefficient. There are other cases in which the solubility decreases as function of temperature (negative coefficient of solubility) or changes very little. For example, in the system 2.1 Na_2Cl_2-H_2O halite has a small positive solubility coefficient whereas that of sylvite in the system 2.2 K_2Cl_2-H_2O is much higher. In the system 2.6 $MgSO_4$-H_2O epsomite and hexahydrite have a positive coefficient of solubility. That of kieserite is negative.

II.3 The Ternary System AX-BX-H_2O

II.3.1 Phase Diagrams

Figure 2.3 shows phase relationships for the case that no solids other than AX_s and BX_s will be forming. The T-AX and T-BX planes represent the boundary systems AX-H_2O and BX-H_2O. The points 1, 2, and 3 correspond to those in Fig. 2.2, and points 4 and 5 are analogous points with respect to the system BX-H_2O. On the surface 1-2-6-4 ice is stable with solutions containing dissolved AX and BX.

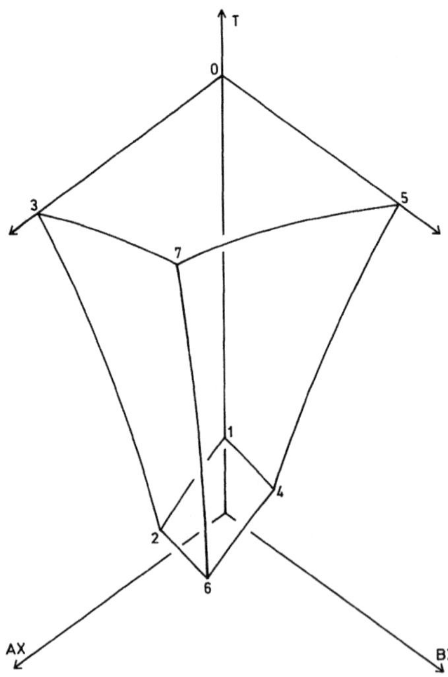

Fig. 2.3 System AX-BX-H_2O. Points 1, 2, and 3 correspond to those in Fig. 2.2. Points 4 and 5 are analogous points in the system BX-H_2O. 1-2-6-4: ice + solution. 2-3-7-6: AX_s + solution. 4-5-7-6: BX_s + solution. 2-6: ice + AX_s + solution. 4-6: ice + BX_s + solution. 6-7: AX_s + BX_s + solution. 6: ice + AX_s + BX_s + solution. 1-2-6-4-5-0-3-7: unsaturated solutions

Surfaces 6-7-3-2 and 6-7-5-4 represent equilibria between saturated solutions and AX_s or BX_s, respectively. Along the line 6-7 saturated solutions are in equilibrium with AX_s and BX_s. The space 1-2-6-4-5-0-3-7 comprises compositions of unsaturated solutions.

In order to represent the phase relationships in a simpler way it is appropriate to project the planes 6-7-3-2 and 6-7-5-4 along the AX and BX axes (Fig. 2.4). The compositions of the solutions are expressed by the relative concentrations

$$f_A = \frac{\{AX\}}{C_T} = \frac{\{A^+\}}{C_T} \tag{II.3.1}$$

and

$$f_B = \frac{\{BX\}}{C_T} = \frac{\{B^+\}}{C_T} \tag{II.3.2}$$

where

$$C_T = \{AX\} + \{BX\} = \{A^+\} + \{B^+\} = \{X^-\} \tag{II.3.3}$$

Figure 2.5 shows an isothermal section of Fig. 2.3 and its projection onto AX-BX.

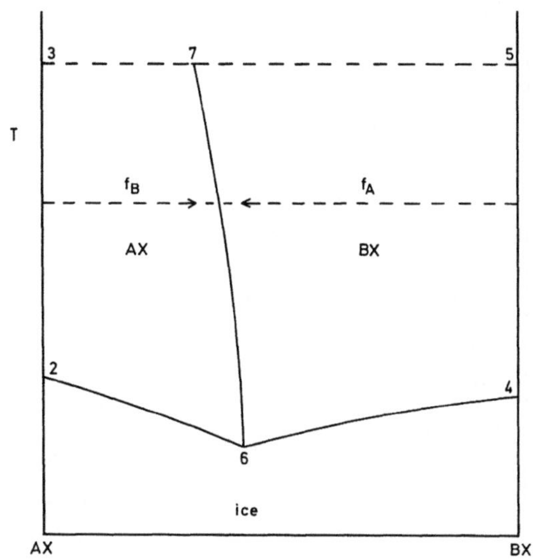

Fig. 2.4 System AX-BX-H_2O. Projection of Fig. 2.3 along the AX and BX axes. Points 2 to 7 correspond to those of Fig. 2.3. f_A, f_B: ion fractions

Instead of Fig. 2.5 a Gibbs triangle may be used (Fig. 2.6). If the concentrations are given in moles dissolved salt/10^3 mol H_2O the relative compositions are expressed as

$$F_{AX} = \frac{\{AX\}}{\{AX\}+\{BX\}+10^3} \tag{II.3.4}$$

$$F_{BX} = \frac{\{BX\}}{\{AX\}+\{BX\}+10^3} \tag{II.3.5}$$

and

$$F_{H_2O} = \frac{10^3}{\{AX\}+\{BX\}+10^3} \tag{II.3.6}$$

The area H_2O-3-7-5 (Fig. 2.6) comprises unsaturated solutions. Point p in the triangle AX-3-7 indicates the overall composition of mixtures of solid AX and a solution which is saturated with respect to AX. The composition of the solution is given by the intersection of line 3-7 with the extension of line AX-p. The same applies to points within the triangle BX-5-7. A point in the triangle AX-7-BX indicates the overall composition of mixtures of solid AX, BX, and solu-

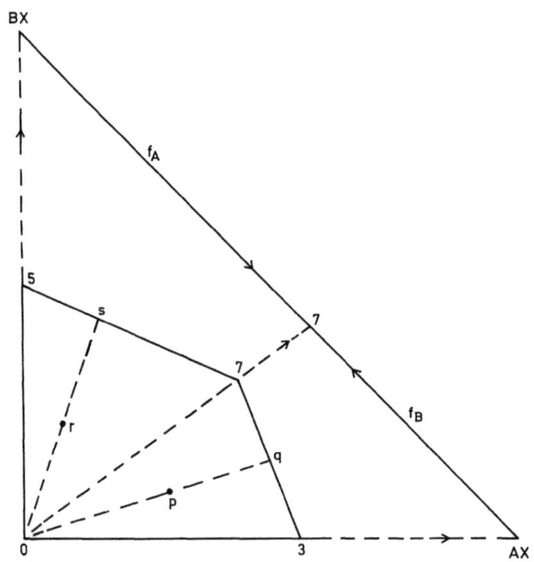

Fig. 2.5 System AX-BX-H_2O. Isothermal section of Fig. 2.3 at the level 0-5-7-3, and projection of the saturation line 3-7-5. Lines p-q-7 and r-s-7 indicate paths of evaporation and crystallization. Points p and r: unsaturated solutions

tion 7. The relative compositions of points projected on AX-BX is given again by Eqs. II.3.1 and II.3.2. It may be noted that the meaning of projected points may be ambiguous, e. g. $f_A = 1$ may represent either an unsaturated solution, the saturated solution 3, a mixture of solid AX and solution 3, or the composition of the solid.

Projections on AX-BX are most useful to show the variation of the total concentration as a function of the composition (Fig. 2.7). Instead of the total concentration, C_T, the so called water value may be used. By convention this value represents the number of moles of water which dissolve 100 mol of water-free solids. The water value, W_T, is related to C_T (moles dissolved salts/1000 mol H_2O) by the equation:

$$W_T = \frac{10^5}{C_T} \tag{II.3.7}$$

II.3.2 Isothermal Evaporation and Dissolution

In Fig. 2.5 point 7 is the endpoint of crystallization. Upon evaporation of water from the undersaturated solution p the concentration increases along p-q until at point q saturation with respect to AX is attained. Upon further evaporation AX_s will be formed. This causes a decrease of the AX/BX ratio of the solution, and the composition of the solution changes along the line q-7. At point 7 BX_s

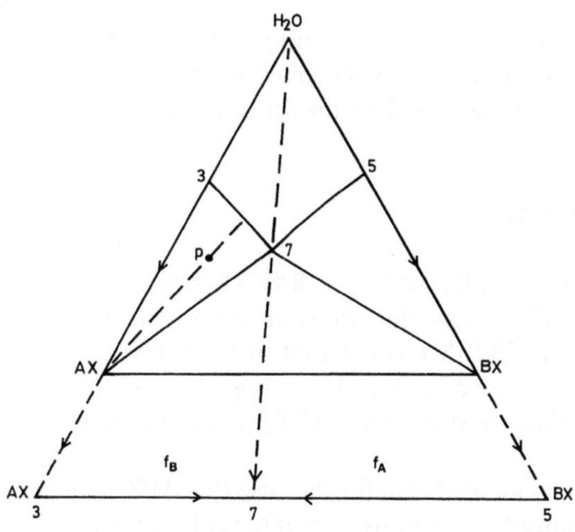

Fig. 2.6 System AX-BX-H_2O. Isotherm at the level 0-5-7-3 of Fig. 2.3 in terms of a Gibbs triangle. H_2O-3-7-5: unsaturated solutions. 3-7-5: saturated solutions. AX-3-7: mixtures of saturated solutions and AX_s. BX-5-7: mixtures of saturated solutions and BX_s. AX-BX-7: mixtures of saturated solution 7, AX_s and BX_s. f_A, f_B: ion fractions. Point p: solid AX + saturated solution

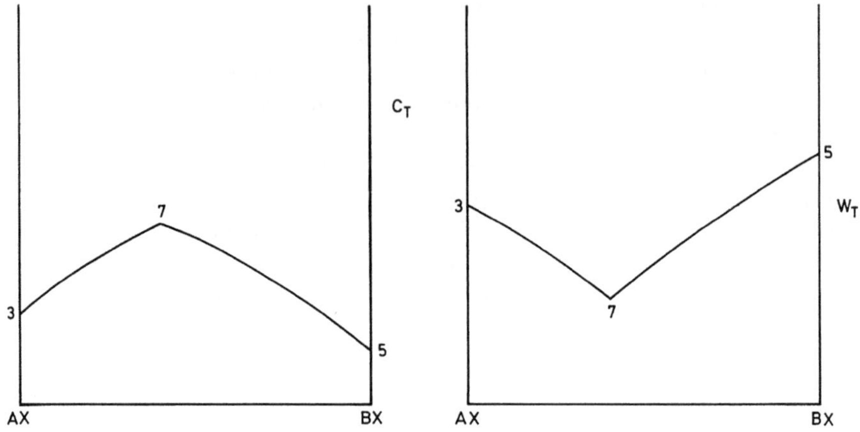

Fig. 2.7 System AX-BX-H_2O. Isothermal phase relations in terms of relative compositions and the total concentration, C_T, or the water value W_T. Undersaturated solutions are located below the line 3-7-5 on the AX-BX-C_T diagram and above this line on the AX-BX-W_T diagram

becomes stable so that upon further evaporation both AX_s and BX_s are precipitated simultaneously. If all water is withdrawn the AX_s/BX_s ratio will be identical with the AX/BX ratio of the initial solution. Similarly, the evaporation of water from solution r leads to the formation of BX_s along the line s-7. If a solution on line 0-7 is evaporated AX_s and BX_s will crystallize simultaneously at point 7.

Upon the addition of water to a mixture of AX_s and BX_s which has the AX/BX ratio of point p the concentration of the solution will evolve along 0-p-q. If the AX/BX ratio of the mixture is that of point s the concentration increases along 0-r-s.

II.3.3
Formation of a Compound

Figure 2.8 shows that solid ABX_2 (AX·BX) is formed with decreasing temperature. Below T_2 $ABX_{2(s)}$ is stable with solutions of various compositions which comprise the same AX/BX ratio as that of the solid. This case is called "congruent solubility". Above T_2 there is "incongruent solubility" because none of the solutions which are stable with $ABX_{2(s)}$ have the same AX/BX ratio as the solid.

Figure 2.9 shows an isotherm for the case that $ABX_{2(s)}$ dissolves congruently. If $ABX_{2(s)}$ is dissolved the composition of the solution evolves along the line 0-3. Vice versa, $ABX_{2(s)}$ will be formed if solution 3 is evaporated. A solution within the triangle 0-2-3 attains saturation with respect to $ABX_{2(s)}$ at a point on 2-3. As $ABX_{2(s)}$ is precipitated the composition of the solution changes towards point 2 where AX_s is formed additionally. If the solution lies within the triangle 0-1-2 sat-

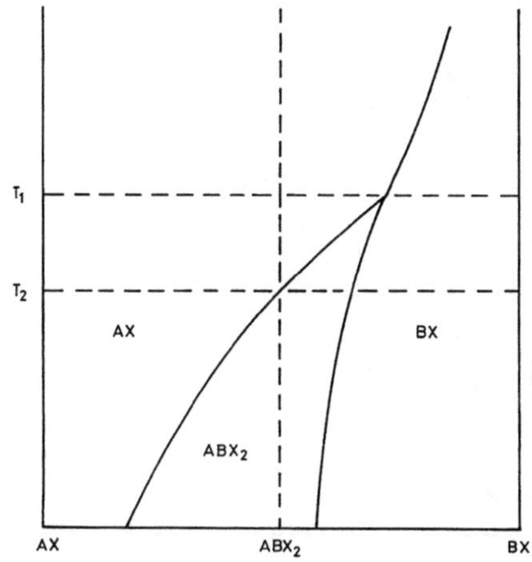

Fig. 2.8 System AX-BX-H$_2$O. Polythermal phase relations. The solid lines denote relative compositions of saturated solutions which are stable with solid AX + ABX$_2$, BX + ABX$_2$, and AX + BX. Between T$_1$ and T$_2$ the compound ABX$_2$ dissolves incongruently. Below T$_2$ it dissolves congruently.

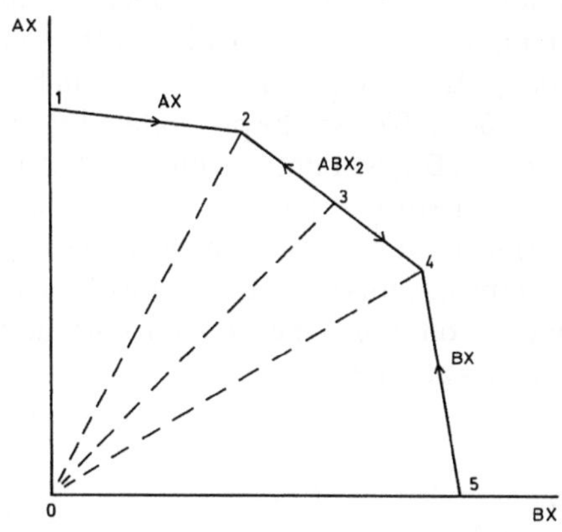

Fig. 2.9 System AX-BX-H$_2$O. Isothermal dissolution and evaporation. ABX$_2$ dissolves congruently. 2-3-4: solutions saturated with respect to ABX$_2$. The slope of the line 0-3 is AX/BX = 1. The arrows indicate crystallization paths

uration with respect to AX$_s$ is attained at first. Upon evaporation of solutions within the triangles 0-3-4 and 0-4-5 point 4 will be reached.

In Fig. 2.10 ABX$_{2(s)}$ dissolves incongruently because at point 4 the AX/BX ratio is higher than that of the solid which is given by the slope of the line 0-3 (AX/BX = 1). Point 2 is the endpoint of crystallization. This point will be reached if solutions within the triangles 0-1-2 and 0-2-4 are evaporated. Evaporating solution 3 leads to crystallization of BX$_s$ along 3-4. At point 4 ABX$_{2(s)}$ becomes stable so

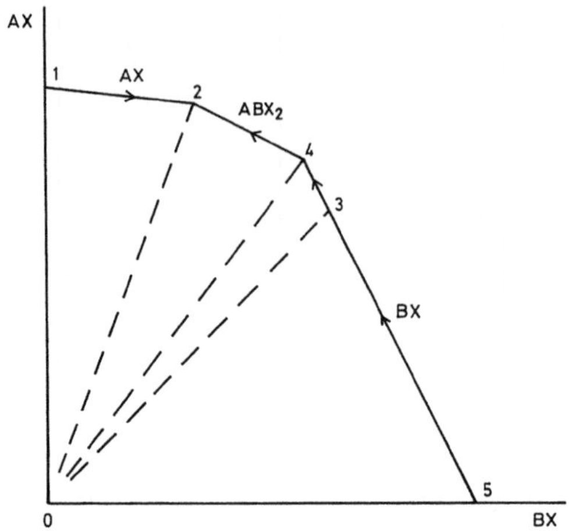

Fig. 2.10 System AX-BX-H_2O. Isothermal dissolution and evaporation. ABX_2 dissolves incongruently. The arrows indicate crystallization paths

that the previously precipitated BX_s reacts with the solution forming $ABX_{2(s)}$. The same happens if solutions within the triangle 0-3-4 are evaporated. However, as their initial AX/BX ratios are >1 the composition of the solution changes along 4-2, precipitating $ABX_{2(s)}$. At point 2 additional AX_s will be formed. From solutions within the triangle 0-3-5 BX_s will be precipitated at first. However, as their initial AX/BX ratios are <1, BX_s will not be consumed completely at point 4 so that the final solids are a mixture of BX_s and $ABX_{2(s)}$. If an excess of $ABX_{2(s)}$ is dissolved the composition of the solution changes along 0-3, attaining saturation with respect to BX_s at point 3. At this point BX_s begins to precipitate, and the solution changes its composition along the line 3-4 until at point 4 equilibrium between BX_s and $ABX_{2(s)}$ is attained.

II.4
Quaternary Systems

II.4.1
Systems with a Common Ion

Figure 2.11 shows a three-dimensional view of an isotherm of the system AX-BX-CX-H_2O. Points 4, 5, and 6 represent equilibria between two solids and solutions of the bounding ternary systems. On the planes 1-4-7-6, 2-4-7-5, and 3-6-7-5, AX, BX, and CX are in equilibrium with saturated quaternary solutions, and along the lines 4-7, 5-7, and 6-7, AX + BX, BX + CX, and AX + CX are stable, respectively. These lines merge into point 7 at which AX + BX + CX is in equilibrium with a defined quaternary solution. Unsaturated solutions are located within the space below the saturation planes.

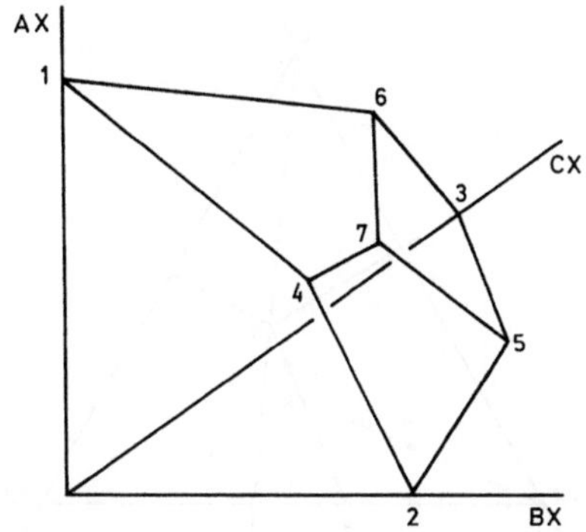

Fig. 2.11 System AX-BX-CX-H_2O. Isothermal equilibria between solids and saturated solutions. 1-4-7-6: AX_s. 2-4-7-5: BX_s. 3-6-7-5: CX_s. 4-7: $AX_s + BX_s$. 5-7: $BX_s + CX_s$. 6-7: $AX_s + CX_s$. 7: $AX_s + BX_s + CX_s$. Space below the planes: unsaturated solutions

A better overview is given if the phase relationships at saturation (Fig. 2.11) are projected along the AX, BX, and CX axes onto an equi-lateral triangle (Fig. 2.12). The relative compositions are given by the expressions

$$f_A = \frac{\{AX\}}{C_T} = \frac{\{A^+\}}{C_T} \qquad \text{(II.4.1)}$$

$$f_B = \frac{\{BX\}}{C_T} = \frac{\{B^+\}}{C_T} \qquad \text{(II.4.2)}$$

and

$$f_C = \frac{\{CX\}}{C_T} = \frac{\{C^+\}}{C_T} \qquad \text{(II.4.3)}$$

where

$$C_T = \{AX\} + \{BX\} + \{CX\} = \{A^+\} + \{B^+\} + \{C^+\} = \{X^-\} \qquad \text{(II.4.4)}$$

The water value is

$$W_T = \frac{10^5}{C_T} \qquad \text{(II.4.5)}$$

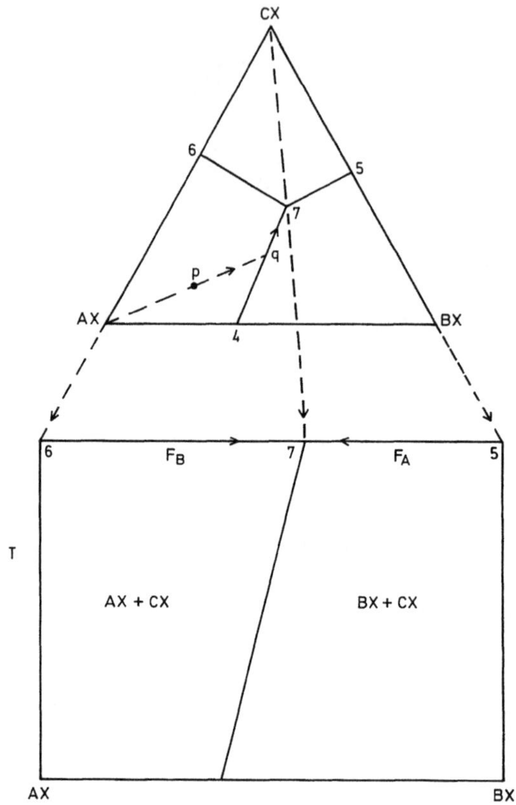

Fig. 2.12 System AX-BX-CX-H_2O. Upper diagram: projection of the planes of Fig. 2.11. The arrows indicate crystallization paths. Lower diagram: projection of the phase relations at saturation with respect to CX_S. F_A, F_B: ion fractions. Both diagrams give relative compositions of saturated solutions which are in equilibrium with solids

The arrows in the triangle of Fig. 2.12 indicate crystallization paths. Solution p is supposed to be saturated with respect to AX. Upon evaporation AX_s will crystallize along the line p-q. Along q-7 AX_s and BX_s are precipitated simultaneously, and at point 7 additional CX_s will be formed.

In Fig. 2.13 the compound ABX_2 is stable. As it dissolves congruently the evaporation of any solution on the line CX-ABX_2 leads to point 3. Upon the evaporation of solutions comprised within AX-CX-ABX_2 or BX-CX-ABX_2 points 1 or 2 will be reached, respectively.

If ABX_2 dissolves incongruently there is a single endpoint of crystallization (point 2, Fig. 2.14). Point 1 is a transition point at which previously precipitated AX_s may be consumed partially or completely, depending on the composition of the initial solution.

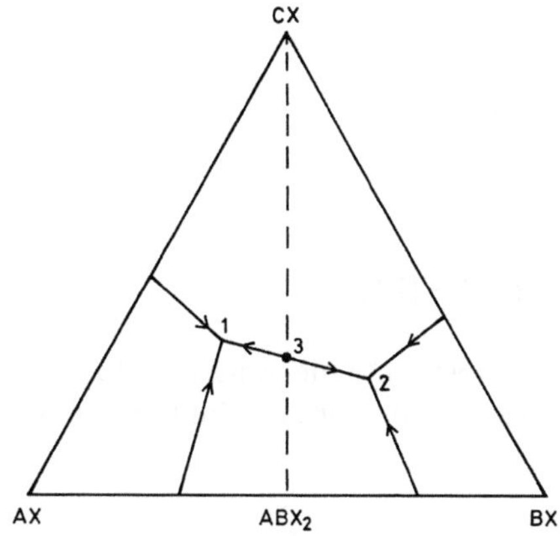

Fig. 2.13 System AX-BX-CX-H_2O. Isothermal phase relations. The arrows indicate crystallization paths. ABX_2 dissolves congruently

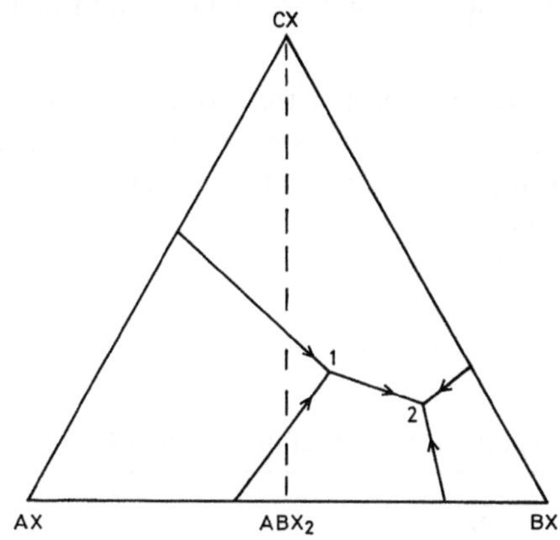

Fig. 2.14 System AX-BX-CX-H_2O. Isothermal phase relations. The arrows indicate crystallization paths. ABX_2 dissolves incongruently

Presenting phase relationships as a function of temperature requires that the system must be viewed at saturation with respect to one of the solids. If in Fig. 2.12 the stabilities of AX_s and BX_s are considered the line 6-7-5 must be projected on AX-BX (saturation with respect to CX). Within the divariant fields of the projection either AX + CX or BX + CX are in equilibrium with saturated solutions. Along the monovariant line AX + BX + CX is stable. Any point in the diagram is expressed by relative concentrations according to the equations

$$F_A = \frac{f_A}{f_A + f_B} = \frac{\{AX\}}{C_P} = \frac{\{A^+\}}{C_P} \qquad \text{(II.4.6)}$$

and

$$F_B = \frac{f_B}{f_A + f_B} = \frac{\{BX\}}{C_P} = \frac{\{B^+\}}{C_P} \quad \text{(II.4.7)}$$

where

$$C_P = \{AX\} + \{BX\} = \{A^+\} + \{B^+\} \quad \text{(II.4.8)}$$

is the concentration with respect to the projection. In order to calculate all concentrations from relative compositions in addition to F_A and F_B the fraction

$$F_C = \frac{\{CX\}}{C_P} = \frac{\{C^+\}}{C_P} \quad \text{(II.4.9)}$$

or the total concentration, C_T, must be known so that $\{C^+\}$ can be calculated from Eqs. II.4.4 or II.4.9. Analogous equations may be obtained for saturation with respect to AX_s or BX_s.

It should be noted that the compositions of solids must be projected analogously to the compositions of solutions according to the above equations. Table 2.1 summarizes the salt points of a number of hypothetical compounds of the system AX-BX-CX-H_2O.

Table 2.1 Salt points of hypothetical solids of the system AX-BX-CX-H_2O

	f_A	f_B	f_C
$A_3BX_4{\cdot}H_2O$	3/4	1/4	
AC_4X_5	1/5		4/5
$BCX_2{\cdot}2H_2O$	1/2	1/2	
$A_5B_3C_2X_{10}$	5/10	3/10	2/10

Saturation with respect to

	AX		BX		CX	
	F_B	F_C	F_A	F_C	F_A	F_B
$A_3BX_4{\cdot}H_2O$	1		1		3/4	1/4
AC_4X_5		1	1/5	4/5	1	
$BCX_2{\cdot}2H_2O$	1/2	1/2		1		1
$A_5B_3C_2X_{10}$	3/5	2/5	5/7	2/7	5/8	3/8

II.4.2 Reciprocal Salt Pairs

II.4.2.1 Phase Diagrams

In a quaternary system with two different cations and anions the exchange reaction $AX_s + BY_s = AY_s + BX_s$ may occur under specific conditions. It should be noted that this reaction is not a solid state reaction but proceeds via dissolution and re-precipitation.

Proving that a reciprocal salt pair represents a quaternary system may be done in the following way. From the equation

$$C_T = \{A^+\} + \{B^+\} = \{X^-\} + \{Y^-\} \tag{II.4.10}$$

where the ion concentrations are defined by the expressions

$$\{A^+\} = \{AX\} + \{AY\} \tag{II.4.11}$$

$$\{B^+\} = \{BX\} + \{BY\} \tag{II.4.12}$$

$$\{X^-\} = \{AX\} + \{BX\} \tag{II.4.13}$$

and

$$\{Y^-\} = \{AY\} + \{BY\} \tag{II.4.14}$$

it follows that if any three of the ions in Eq. II.4.10 are chosen the fourth ion cannot be varied independently, e. g. choosing $\{A^+\}$, $\{B^+\}$, and $\{X^-\}$ implies that $\{Y^-\}$ is defined (electroneutrality). Thus the components are A^+, B^+, X^-, and H_2O.

It is most suitable to present a reciprocal salt pair at constant temperature by a four sided pyramid (Fig. 2.15). The triangles on the outside correspond to the ternary systems AX-BX-H_2O, BX-BY-H_2O, BY-AY-H_2O, and AY-AX-H_2O according to Fig. 2.6. Unsaturated and saturated solutions are located above and on the planes, respectively.

Projecting Fig. 2.15 from the H_2O point along the edges yields Fig. 2.16. The arrows indicate crystallization paths. A mixture of solid AX and BY dissolves congruently, and the evaporation of solution 7 leads to the formation of AX_s and BY_s. Point 7 is a pseudo ternary point. If a mixture of solid AY and BX is dissolved the above exchange reaction proceeds. In Fig. 2.17 a mixture of solid AX and BY dissolves incongruently. Point 6 is the endpoint of crystallization whereas point 5 is a transition point. For example, a solution located on line 1-5 and within the triangle AX-BX-BY will precipitate AX + AY upon evaporation until point 5 is attained. At this point BY is produced while AY is consumed. Subsequently, the solution changes its composition along the line 5-6 precipitating AX + BY. Evaporating to dryness at point 6 yields AX + BX + BY.

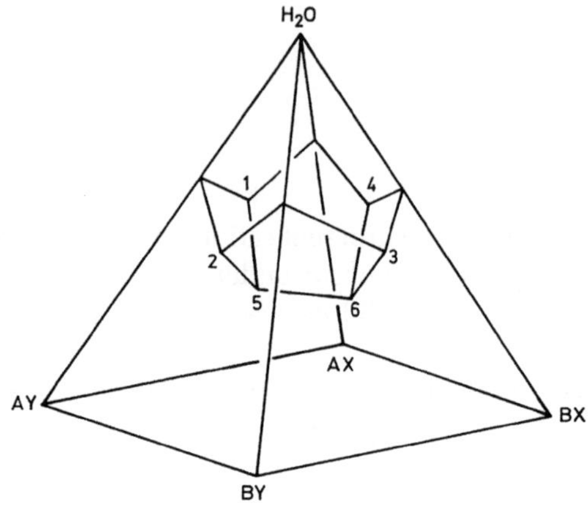

Fig. 2.15 System AX-BX-AY-BY-H_2O (reciprocal salt pair AX + BY = AY + BX). Isothermal equilibria between solids and saturated solutions. Boundary systems according to Fig. 2.6. 1-5: $AX_s + AY_s$. 2-5: $AY_s + BY_s$. 3-6: $BY_s + BX_s$. 4-6: $BX_s + AX_s$. 5-6: $AX_s + BY_s$. 5: $AX_s + AY_s + BY_s$. 6: $AX_s + BX_s + BY_s$

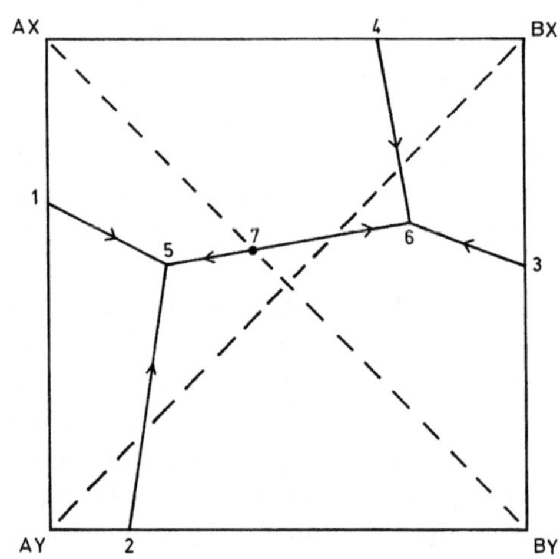

Fig. 2.16 System AX-BX-AY-BY-H_2O (reciprocal salt pair AX + BY = AY + BX). Projection of isothermal phase relations. The arrows indicate crystallization paths. A mixture of solid AX and BY dissolves congruently. Point 7 is pseudo ternary

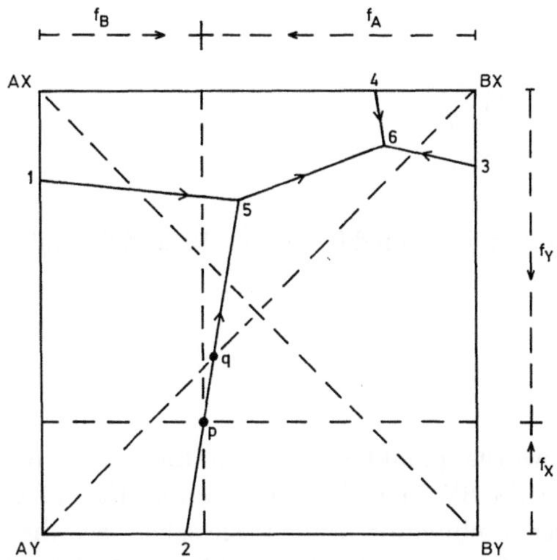

Fig. 2.17 System AX-BX-AY-BY-H_2O (reciprocal salt pair AX + BY = AY + BX). Projection of isothermal phase relations. The arrows indicate crystallization paths. A mixture of solid AX and BY dissolves incongruently. f_A, f_B, f_X, f_Y: ion fractions

II.4.2.2 Concentrations

The relative composition of point p in Fig. 2.17 is given by the fractions of the four ions according to the expressions

$$f_A = \frac{\{A^+\}}{\{A^+\}+\{B^+\}} = \frac{\{A^+\}}{C_T} \tag{II.4.15}$$

$$f_B = \frac{\{B^+\}}{\{A^+\}+\{B^+\}} = \frac{\{B^+\}}{C_T} \tag{II.4.16}$$

$$f_X = \frac{\{X^-\}}{\{X^-\}+\{Y^-\}} = \frac{\{X^-\}}{C_T} \tag{II.4.17}$$

and

$$f_Y = \frac{\{Y^-\}}{\{X^-\}+\{Y^-\}} = \frac{\{Y^-\}}{C_T} \qquad \text{(II.4.18)}$$

The total concentration is defined by Eq. II.4.10, and the water value is

$$W_T = \frac{10^5}{C_T} \qquad \text{(II.4.19)}$$

Figure 2.17 shows that point p is located in the triangle AX-BY-AY as well as in the triangle AY-BX-BY. Therefore, the composition of this point may be expressed by either combination of the three salts. Nevertheless it may be also expressed in terms of the triangles AX-BY-BX and AY-BX-AX. The concentrations of the various combinations are interrelated. If one set of concentrations is known the others may be calculated. Table 2.2 summarizes the four combinations and expressions which are obtained by writing Eqs. II.4.11-II.4.14 for each of the cases.

It should be pointed out that in two of the four cases in Table 2.2 all concentrations have positive values, whereas in each of the other two cases one of the concentrations will have a negative value. For example, with respect to point p (Fig. 2.17) the ion fractions decrease in the order $f_Y > f_A > f_B > f_X$ so that according to Eqs. II.4.15-II.4.18 $\{Y^-\} > \{A^+\} > \{B^+\} > \{X^-\}$. It follows that in cases 3 and 4 (Table 2.2) all concentrations have positive values. However, in cases 1 and 2 the concentrations of AX and BX are $\{AX\} = \{A^+\} - \{Y^-\} = \{X^-\} - \{B^+\} < 0$ and $\{BX\} = \{B^+\} - \{Y^-\} = \{X^-\} - \{A^+\} < 0$.

Negative values of concentrations become important if the composition of a solution is traced. For example, if the compositions of solutions along the path p-q (Fig. 2.17) can be described by two functions, {BY} = func.{AY} and {BX} = func.{AY}, it is desirable to use these same functions in order to express the compositions of the solutions along the path q-5, although {BY} = func.{AY} < 0.

II.4.2.3 Polytherms

In order to pursue phase relationships as a function of temperature it is required to view the system at saturation with respect to one of the solids. Figure 2.18 shows phase relationships at saturation with respect to AX. Point p is an invariant point at which four solids are stable. At temperatures above and below this point AX + BY and AY + BX are the stable salt pairs, respectively (AX + BY at T_2, upper square, AY + BX at T_1, lower square).

Table 2.2 System AX-BX-AY-BY-H_2O (reciprocal salt pair AX + BY = AY + BX). Combination of concentrations and pertinent equations for calculation

Combinations

	{AX}	{BX}	{AY}	{BY}
1.	+	+	+	0
2.	+	+	0	+
3.	+	0	+	+
4.	0	+	+	+

Expressions for concentrations

1. $\{BY\} = 0$
   $\{BX\} = \{B^+\}, \{AY\} = \{Y^-\}, \{AX\} = \{A^+\} - \{Y^-\} = \{X^-\} - \{B^+\}$

2. $\{AY\} = 0$
   $\{AX\} = \{A^+\}, \{BY\} = \{Y^-\}, \{BX\} = \{B^+\} - \{Y^-\} = \{X^-\} - \{A^+\}$

3. $\{BX\} = 0$
   $\{BY\} = \{B^+\}, \{AX\} = \{X^-\}, \{AY\} = \{Y^-\} - \{B^+\} = \{A^+\} - \{X^-\}$

4. $\{AX\} = 0$
   $\{AY\} = \{A^+\}, \{BX\} = \{X^-\}, \{BY\} = \{Y^-\} - \{A^+\} = \{B^+\} - \{X^-\}$

At saturation with respect to AX the relative compositions of solutions are expressed by the equations

$$F_B = \frac{f_B}{f_B + f_Y} = \frac{\{B^+\}}{C_P} \tag{II.4.20}$$

and

$$F_Y = \frac{f_Y}{f_B + f_Y} = \frac{\{Y^-\}}{C_P} \tag{II.4.21}$$

where $\{B^+\}$ and $\{Y^-\}$ are defined by equations II.4.12 and II.4.14 and

$$C_P = \{B^+\} + \{Y^-\} \tag{II.4.22}$$

is the concentration with respect to the projection. The water value is given by the equation

$$W_P = \frac{10^5}{C_P} \tag{II.4.23}$$

In order to calculate all concentrations from relative compositions, in addition to F_B (or F_Y) and C_P (or W_P) either the fraction

$$F_A = \frac{\{A^+\}}{C_P} \tag{II.4.24}$$

or the fraction

$$F_X = \frac{\{X^-\}}{C_P} \tag{II.4.25}$$

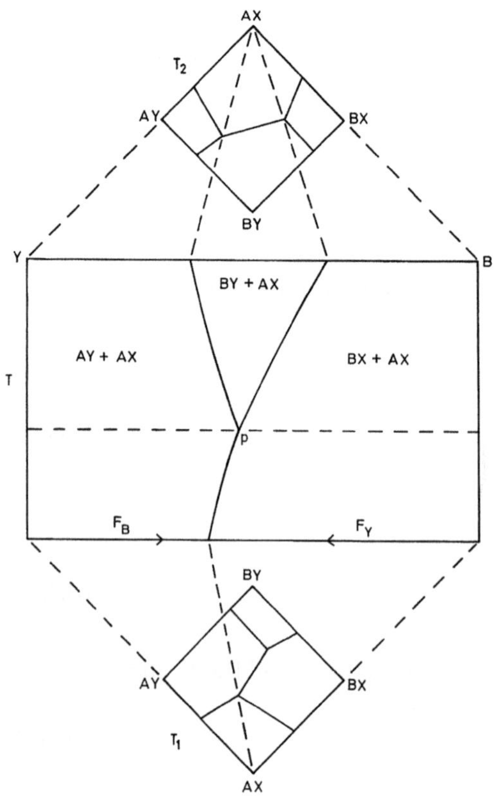

Fig. 2.18 System AX-BX-AY-BY-H_2O (reciprocal salt pair AX + BY = AY + BX). Projection of phase relationships at saturation with respect to AX as a function of temperature. At point p a saturated solution is stable with solid AX, BX, AY, and BY. The upper and lower horizontal lines of the rectangle correspond to the projections of the squares at T_2 and T_1

must be known so that either $\{A^+\}$ or $\{X^-\}$ can be calculated from Eq. II.4.10. If both F_X and F_A are not given the total concentration, C_T, must be known so that $\{A^+\}$ and $\{X^-\}$ can be obtained from the expressions

$$\{A^+\} = C_T - C_P + \{Y^-\} \tag{II.4.26}$$

and

$$\{X^-\} = C_T - C_P + \{B^+\} \tag{II.4.27}$$

Analogous equations may be derived if projections from AY, BX, or BY are considered. For example, projecting from BY yields the expressions

$$F_A = \frac{f_A}{f_A + f_X} = \frac{\{A^+\}}{C_P} \tag{II.4.28}$$

and

$$F_X = \frac{f_X}{f_A + f_X} = \frac{\{X^-\}}{C_P} \tag{II.4.29}$$

where

$$C_P = \{A^+\} + \{X^-\} \tag{II.4.30}$$

The water value is

$$W_P = \frac{10^5}{C_P} \tag{II.4.31}$$

The concentrations of B^+ and Y^- may be obtained from the fraction

$$F_B = \frac{\{B^+\}}{C_P} \tag{II.4.32}$$

$$F_Y = \frac{\{Y^-\}}{C_P} \tag{II.4.33}$$

or from the expressions

$$\{B^+\} = C_T - C_P + \{X^-\} \tag{II.4.34}$$

Table 2.3 Salt points of hypothetical compounds of the system AX-BX-AY-BY-H_2O (reciprocal salt pair AX + BY = AY + BX)

	f_A	f_B	f_X	f_Y
$A_5XY_4 \cdot 2H_2O$	1		1/5	4/5
AB_3X_4	1/4	3/4	1	
$A_4BX_3Y_2 \cdot H_2O$	4/5	1/5	3/5	2/5

Saturation with respect to

	AX		BY	
	F_B	F_Y	F_A	F_X
$A_5XY_4 \cdot 2H_2O$		1	5/6	1/6
AB_3X_4	1		1/5	4/5
$A_4BX_3Y_2 \cdot H_2O$	1/3	2/3	4/7	3/7

and

$$\{Y^-\} = C_T - C_P + \{A^+\} \tag{II.4.35}$$

Salt points are expressed according to the ion fractions (Eqs. II.4.15-II.4.18, II.4.20 and II.4.21). Table 2.3 summarizes some points for hypothetical solids.

II.5 The Quinary System AX-BX-CX-AY-BY-CY-H_2O

II.5.1 Isotherms

Figure 2.19 shows the construction for an isotherm of the quinary system AX-BX-CX-AY-BY-CY-H_2O. The surface of the prism is given by the projections of the reciprocal salt pairs AX + BY = AY + BX, BX + CY = BY + CX, CX + AY = CY + AX, and the projections of the quaternary systems AX-BX-CX-H_2O and AY-BY-CY-H_2O. A point within the prism represents a quinary solution.

Figure 2.20 shows the phase relationships of a system in which no solids other than AX, BX, CX, AY, BY, and CY occur. Within the prism there are twelve planes at which two solids are in equilibrium with saturated solutions. Along the ten lines three solids are stable, and at the three isothermal invariant points saturated solutions are in equilibrium with four solids.

II.5.2
Concentrations

The total concentration of a quinary solution is given by the equation

$$C_T = \{A^+\} + \{B^+\} + \{C^+\} = \{X^-\} + \{Y^-\} \tag{II.5.1}$$

where the ion concentrations are defined by the expressions

$$\{A^+\} = \{AX\} + \{AY\} \tag{II.5.2}$$
$$\{B^+\} = \{BX\} + \{BY\} \tag{II.5.3}$$
$$\{C^+\} = \{CX\} + \{CY\} \tag{II.5.4}$$
$$\{X^-\} = \{AX\} + \{BX\} + \{CX\} \tag{II.5.5}$$

and

$$\{Y^-\} = \{AY\} + \{BY\} + \{CY\}. \tag{II.5.6}$$

The water value is

$$W_T = \frac{10^5}{C_T} \tag{II.5.7}$$

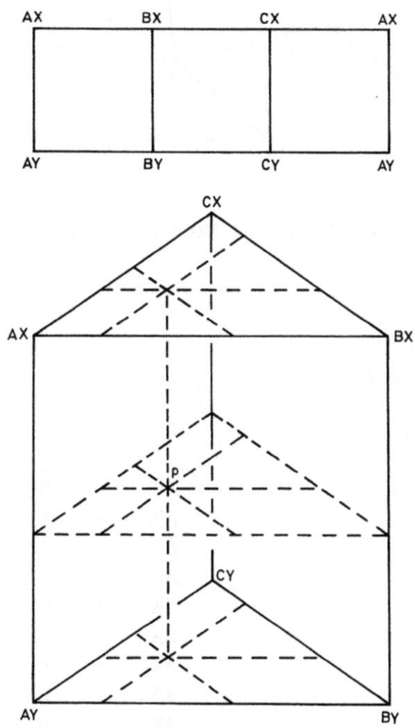

Fig. 2.19 System AX-BX-CX-AY-BY-CY-H_2O. Construction of an isotherm in terms of the ion fractions $f_A + f_B + f_C = 1$ and $f_X + f_Y = 1$. The triangles are projections of the quaternary system AX-BX-CX-H_2O and AY-BY-CY-H_2O. The squares are the projections of the reciprocal salt pairs AX + BY = AY + BX, AX + CY = AY + CX, and BX + CY = BY + CX

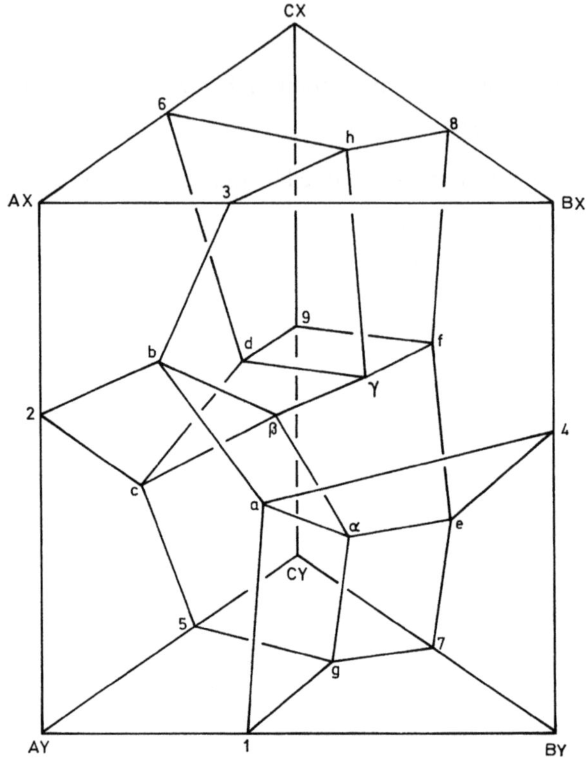

Fig. 2.20 System AX-BX-CX-AY-BY-CY-H_2O. Isothermal equilibria between saturated solutions and solids.

AX_s:	AX-3-b-2-c-d-6-h-γ-ß
BX_s:	BX-8-h-3-b-a-4-e-f-γ-ß-α
CX_s:	CX-6-h-8-f-γ-d-9
AY_s:	AY-2-b-a-1-g-α-ß-c-5
BY_s:	BY-1-a-4-e-α-g-7
CY_s:	CY-5-g-7-e-f-9-d-c-ß-γ

AX_s	+	BX_s:	3-h-γ-ß-b
AX_s	+	CX_s:	6-h-γ-d
AX_s	+	AY_s:	2-b-ß-c
AX_s	+	CY_s:	c-d-γ-ß
BX_s	+	CX_s:	8-f-γ-h
BX_s	+	BY_s:	4-e-α-a
BX_s	+	CY_s:	α-ß-γ-f-e
CX_s	+	CY_s:	9-f-γ-d
AY_s	+	BY_s:	1-a-α-g
AY_s	+	CY_s:	5-g-α-ß-c
AY_s	+	BX_s:	a-b-ß-α
BY_s	+	CY_s:	7-g-α-e

AY_s	+	BY_s	+	BX_s:	a-α
AX_s	+	AY_s	+	BX_s:	b-ß
AX_s	+	AY_s	+	CY_s:	c-ß
AX_s	+	CX_s	+	CY_s:	d-γ
BX_s	+	BY_s	+	CY_s:	e-α
BX_s	+	CX_s	+	CY_s:	f-γ
AY_s	+	BY_s	+	CY_s:	g-α
AX_s	+	BX_s	+	CX_s:	h-γ
AY_s	+	BX_s	+	CY_s:	α-ß
AX_s	+	BX_s	+	CY_s:	β-γ

AY_s	+	BY_s	+	BX_s	+	CY_s:	α
AX_s	+	AY_s	+	BX_s	+	CY_s:	ß
AX_s	+	BX_s	+	CX_s	+	CY_s:	γ

The relative composition of a solution (e. g. point p, Fig. 2.19) is given by the equations:

$$f_A = \frac{\{A^+\}}{C_T} \quad \text{(II.5.8)}$$

$$f_B = \frac{\{B^+\}}{C_T} \quad \text{(II.5.9)}$$

$$f_C = \frac{\{C^+\}}{C_T} \quad \text{(II.5.10)}$$

$$f_X = \frac{\{X^-\}}{C_T} \quad \text{(II.5.11)}$$

and

$$f_Y = \frac{\{Y^-\}}{C_T} \quad \text{(II.5.12)}$$

If any four of the ions in Eq. II.5.1 are chosen the fifth ion cannot be varied independently. This also implies that any combination of four salts out of six fulfills the requirement of the above equation (electroneutrality). Therefore, the concentrations of four salts are needed to express the composition of a quinary solution. There are twelve combinations of concentrations (Table 2.4). If the concentrations for one of the cases are known the others may be calculated. Table 2.5 summarizes expressions which have been obtained by writing Eqs. II.5.2-II.5.6 for each of the combinations in Table 2.4.

Table 2.4 Combinations of concentrations for the quinary system AX-BX-CX-AY-BY-CY-H_2O

	{AX}	{BX}	{CX}	{AY}	{BY}	{CY}
1.	0	0	+	+	+	+
2.	0	+	0	+	+	+
3.	0	+	+	+	0	+
4.	0	+	+	+	+	0
5.	+	0	0	+	+	+
6.	+	0	+	0	+	+
7.	+	0	+	+	+	0
8.	+	+	0	0	+	+
9.	+	+	0	+	0	+
10.	+	+	+	0	0	+
11.	+	+	+	0	+	0
12.	+	+	+	+	0	0

Table 2.5 Expressions for the calculation of concentrations for the combinations of Table 2.4

1. $\{AX\} = \{BX\} = 0$ $\{CX\} = \{X^-\}, \{AY\} = \{A^+\}, \{BY\} = \{B^+\}, \{CY\} = \{C^+\} - \{X^-\}$
2. $\{AX\} = \{CX\} = 0$ $\{BX\} = \{X^-\}, \{AY\} = \{A^+\}, \{BY\} = \{B^+\} - \{X^-\}, \{CY\} = \{C^+\}$
3. $\{AX\} = \{BY\} = 0$ $\{BX\} = \{B^+\}, \{CX\} = \{X^-\} - \{B^+\}, \{AY\} = \{A^+\}, \{CY\} = \{Y^-\} - \{A^+\}$
4. $\{AX\} = \{CY\} = 0$ $\{BX\} = \{X^-\} - \{C^+\}, \{CX\} = \{C^+\}, \{AY\} = \{A^+\}, \{BY\} = \{Y^-\} - \{A^+\}$
5. $\{BX\} = \{CX\} = 0$ $\{AX\} = \{X^-\}, \{AY\} = \{A^+\} - \{X^-\}, \{BY\} = \{B^+\}, \{CY\} = \{C^+\}$
6. $\{BX\} = \{AY\} = 0$ $\{AX\} = \{A^+\}, \{CX\} = \{X^-\} - \{A^+\}, \{BY\} = \{B^+\}, \{CY\} = \{Y^-\} - \{B^+\}$
7. $\{BX\} = \{CY\} = 0$ $\{AX\} = \{X^-\} - \{C^+\}, \{CX\} = \{C^+\}, \{AY\} = \{Y^-\} - \{B^+\}, \{BY\} = \{B^+\}$
8. $\{CX\} = \{AY\} = 0$ $\{AX\} = \{A^+\}, \{BX\} = \{X^-\} - \{A^+\}, \{BY\} = \{Y^-\} - \{C^+\}, \{CY\} = \{C^+\}$
9. $\{CX\} = \{BY\} = 0$ $\{AX\} = \{X^-\} - \{B^+\}, \{BX\} = \{B^+\}, \{AY\} = \{Y^-\} - \{C^+\}, \{CY\} = \{C^+\}$
10. $\{AY\} = \{BY\} = 0$ $\{AX\} = \{A^+\}, \{BX\} = \{B^+\}, \{CX\} = \{C^+\} - \{Y^-\}, \{CY\} = \{Y^-\}$
11. $\{AY\} = \{CY\} = 0$ $\{AX\} = \{A^+\}, \{BX\} = \{B^+\} - \{Y^-\}, \{CX\} = \{C^+\}, \{BY\} = \{Y^-\}$
12. $\{BY\} = \{CY\} = 0$ $\{AX\} = \{A^+\} - \{Y^-\}, \{BX\} = \{B^+\}, \{CX\} = \{C^+\}, \{AY\} = \{Y^-\}$

II.5.3 Projections of Isotherms

In general the phase relationships of a quinary system are more complicated than those shown in Fig. 2.20 so that the system must be viewed at saturation with respect to one of the solids. In Fig. 2.21 saturation with respect to AX is considered. Projecting along the edges AX-BX, AX-CX, and AX-AY yields the triangle F_B-F_C-F_Y. From Fig. 2.22 it may be seen that two of the sides of the triangle (diagram 2) correspond to the projections of the reciprocal salt pairs AX + BY = AY + BX (diagram 3) and AX + CY = AY + CX (diagram 4). The third side

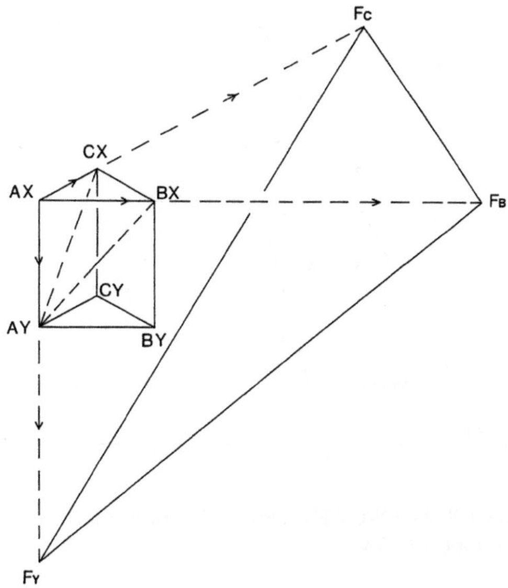

Fig. 2.21 System AX-BX-CX-AY-BY-CY-H_2O. Projection of an isotherm at saturation with respect to AX

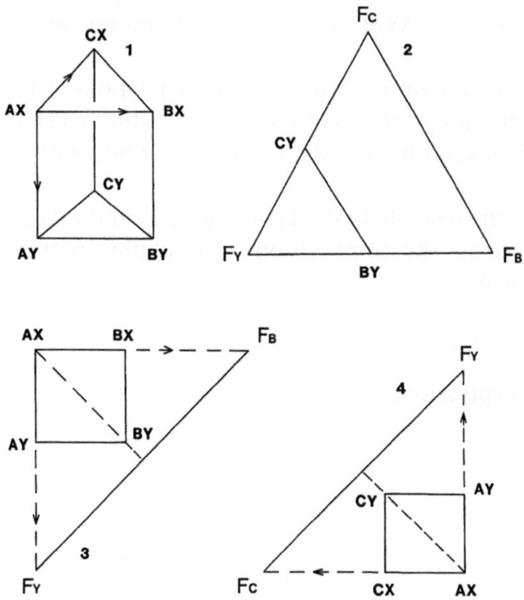

Fig. 2.22 System AX-BX-CX-AY-BY-CY-H_2O. Diagrams 1 and 2: prism and its projection according to Fig. 2.21. Diagrams 3 and 4: projections of the reciprocal salt pairs AX + BY = AY + BX and AX + CY = AY + CX, respectively. The line F_B-F_C in diagram 2 is the projection of the system AX-BX-CX-H_2O at saturation with respect to AX

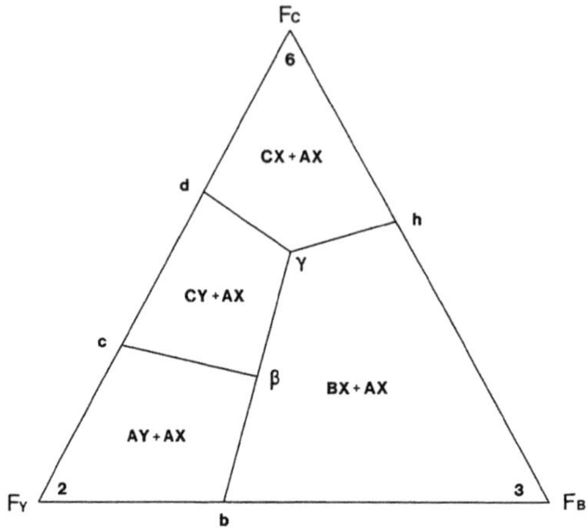

Fig. 2.23 System AX-BX-CX-AY-BY-CY-H_2O. Projection of isothermal phase relations of Fig. 2.20 at saturation with respect to AX

of diagram 2 corresponds to the projection of the quaternary system AX-BX-CX-H_2O. The line BY-CY in diagram 2 divides the triangle into two areas. The trapeze BY-CY-F_C-F_B and the triangle F_Y-CY-BY may be taken as the systems BX-CX-BY-CY-H_2O and AY-BY-CY-H_2O at saturation with respect to AX, respectively.

Figure 2.23 shows the projection of the phase relations of Fig. 2.20. The four fields represent the compositions of quinary AX-saturated solutions which are in equilibrium with solid BX, CX, AY, and CY. Equilibria with respect to BY are not projected.

As the sides of the triangle F_Y-F_B-F_C in Figs. 2.22 and 2.23 are projections of the bounding systems the composition of a solution within the triangle is expressed relative to

$$C_P = \{B^+\} + \{C^+\} + \{Y^-\} \tag{II.5.13}$$

according to the expressions

$$F_B = \frac{\{B^+\}}{C_P} \tag{II.5.14}$$

$$F_C = \frac{\{C^+\}}{C_P} \tag{II.5.15}$$

$$F_Y = \frac{\{Y^-\}}{C_P} \tag{II.5.16}$$

where the ion concentrations are defined by expressions II.5.3, II.5.4, and II.5.6.

In order to calculate concentrations from relative compositions it is necessary to know either C_P or the water value

$$W_P = \frac{10^5}{C_P} \tag{II.5.17}$$

and one of the fractions

$$F_A = \frac{\{A^+\}}{C_P} \tag{II.5.18}$$

$$F_X = \frac{\{X^-\}}{C_P} \tag{II.5.19}$$

so that either $\{X^-\}$ or $\{A^+\}$ may be calculated from Eq. II.5.1. If F_A and F_X are not given the total concentration, C_T, must be known in order to calculate $\{A^+\}$ and $\{X^-\}$ from the expressions

$$\{A^+\} = C_T - C_P + \{Y^-\} \tag{II.5.20}$$

and

$$\{X^-\} = C_T - C_P + \{B^+\} + \{C^+\} \tag{II.5.21}$$

Salt points are expressed according to the projection of solutions. Table 2.6 shows examples for hypothetical solids.

Table 2.6. Salt points of hypothetical compounds of the system AX-BX-CX-AY-BY-CY-H_2O

	Position in Fig. 2.19				
	f_A	f_B	f_C	f_X	f_Y
$A_5XY_4{\cdot}2H_2O$	1			1/5	4/5
AB_3X_4	1/4	3/4		1	
$A_4BX_3Y_2{\cdot}H_2O$	4/5	1/5		3/5	2/5
$ABCX_3{\cdot}4H_2O$	1/3	1/3	1/3	1	
ABC_2Y_4	1/4	1/4	2/4		1
$A_2BC_3X_2Y_4$	2/6	1/6	3/6	2/6	4/6

	Saturation with respect to AX		
	F_B	F_C	F_Y
$A_5XY_4{\cdot}2H_2O$			1
AB_3X_4	1		
$A_4BX_3Y_2{\cdot}H_2O$	1/3		2/3
$ABCX_3{\cdot}4H_2O$	1/2	1/2	
ABC_2Y_4	1/7	2/7	4/7
$A_2BC_3X_2Y_4$	1/8	3/8	4/8

II.5.4
Polytherms

Presenting a quinary system as a function of temperature depends very much on the case and the phase relations which are intended to be shown. With respect to Fig. 2.23 it is most appropriate to project the line b-β-γ-h from the F_B-point onto the line F_Y-F_C so that points β and γ can be plotted as a function of temperature. However, if more complicated systems are considered it may be more convenient to present phase relations as a function of temperature and the total concentration of one of the ions. An example is the quinary seawater system where it is most suitable to use the total concentration of Mg^{2+} (Chapter III. 5).

II.6
Numerical Examples

II.6.1
System 2.1 Na_2Cl_2-H_2O

ha: halite (NaCl)

II.6.1.1
Isothermal evaporation

If a solution 1 of 10 mol Na_2Cl_2 + 1000 mol water ($\{Na_2Cl_2\} = 10$) is evaporated at 25 °C, saturation with respect to halite is attained at a concentration of $\{Na_2Cl_2\} = 55.5$ (solution 2). The quantity of the lost and the remaining water may be calculated from the balances

moles	solution 1		solution 2		vapor
Na_2Cl_2	10	=	$55.5 \cdot u$		
H_2O	10^3	=	$10^3 \cdot u$	+	v

$u = 0.1802, v = 819.8.$

If all the water is evaporated the quantity of the solid will be 10 mol Na_2Cl_2 or 20 mol halite.

II.6.1.2
Dissolution and Precipitation

Stirring at 25 °C a quantity of halite corresponding to 57.9 mol Na_2Cl_2 with 1000 mol of water yields the above equilibrium solution 2 ($\{Na_2Cl_2\} = 55.5$). The quantity of the remaining solid is calculated from the balances

moles	water	halite 1		solution 2		halite 2
Na_2Cl_2		57.9	=	$55.5 \cdot u$	+	v
H_2O	10^3		=	$10^3 \cdot u$		

$u = 1, v = 2.4.$

Upon cooling a saturated solution from 95 to 50 °C halite will be precipitated. Its quantity is calculated from the equations

moles	solution 95 °C		solution 50 °C		halite
Na_2Cl_2	59.9	=	$56.7 \cdot u$	+	v
H_2O	10^3	=	$10^3 \cdot u$		

$u = 1, v = 3.2.$

II.6.2 System 2.6 $MgSO_4$-H_2O

hx: hexahydrite ($MgSO_4 \cdot 6H_2O$)
ks: kieserite ($MgSO_4 \cdot H_2O$)

II.6.2.1 Dissolution of Kieserite

If a saturated solution of 80.3 mol $MgSO_4$ + 1000 mol H_2O and an excess of kieserite of 20 mol are cooled from 85° to 67.5 °C, a portion of the kieserite will be dissolved (negative coefficient of solubility). The quantity of the dissolved kieserite and the composition of the solution is calculated from the mass balances

	85 °C				67.5 °C		
moles	kieserite 1		solution		kieserite 2		solution
$MgSO_4$	20	+	80.3	=	u	+	$87.8 \cdot v$
H_2O	20	+	10^3	=	u	+	$10^3 \cdot v$

$u = 11.8, v = 1.0082.$

II.6.2.2 Precipitation of Hexahydrite

If 10 mol of kieserite and a saturated solution of 87.8 mol $MgSO_4$ + 1000 moles H_2O are cooled from 67.5 to 60 °C hexahydrite will become stable. As the solubility of hexahydrite decreases with decreasing temperature it will be precipitated, and the initial kieserite will dissolve because it is unstable below 67.5 °C.

Thus, $MgSO_4$ and water are simultaneously added to the solution and withdrawn from it via the dissolution of kieserite and the precipitation of hexahydrite, respectively. At 60 °C the concentration of the solution saturated with respect to hexahydrite is $\{MgSO_4\} = 81.5$. The quantity of the precipitated hexahydrite is calculated from the equations

			67.5 °C		60 °C		
moles	kieserite		solution		hexahydrite		solution
$MgSO_4$	10	+	87.8	=	u	+	$81.5 \cdot v$
H_2O	10	+	10^3	=	$6 \cdot u$	+	$10^3 \cdot v$

u = 30.3, v = 0.8282.

It appears that 30.3 mol hexahydrite have been precipitated.

II.6.3
System 3.1 Na_2Cl_2-K_2Cl_2-H_2O

ha: halite (NaCl)
sy: sylvite (KCl)

II.6.3.1
Projections

At 20 °C the compositions of some of the solutions are

solution	solids	$\{Na_2Cl_2\}$	$\{K_2Cl_2\}$	C_T
1	ha	55.4	0	55.4
2	ha	52.4	6.0	58.4
3	ha	49.5	12.0	61.5
4	ha + sy	46.6	18.2	64.8
5	sy	35.0	22.5	57.5
6	sy	15.0	32.2	47.2
7	sy	0	41.5	41.5

The relative compositions of the solutions in terms of ion fractions (Eqs. II.3.1-II.3.3) and in terms of a Gibbs triangle (Eqs. II.3.4-II.3.6) are

solution	solids	f_{Na_2}	f_{K_2}	$F_{Na_2Cl_2}$	$F_{K_2Cl_2}$	F_{H_2O}
1	ha	1	0	0.0525	0	0.9475
2	ha	0.897	0.103	0.0495	0.0057	0.9448
3	ha	0.805	0.195	0.0466	0.0113	0.9421
4	ha + sy	0.719	0.281	0.0438	0.0171	0.9391
5	sy	0.609	0.391	0.0331	0.0213	0.9456
6	sy	0.318	0.682	0.0143	0.0307	0.9550
7	sy	0	1	0	0.0398	0.9602

II.6.3.2
Isothermal Evaporation

It is assumed that an undersaturated solution 1, containing 26.2 mol Na_2Cl_2 + 3.0 mol K_2Cl_2 + 1000 mol water is evaporated at 20 °C. As this solution has the same Na_2Cl_2/K_2Cl_2 ratio as that of the above saturated solution 2, halite will begin to precipitate if a concentration $C_T = 58.4$ is attained. The composition of the solution at this stage is calculated from the equations

moles	solution 1		solution 2		vapor
Na_2Cl_2	26.2	=	$52.4 \cdot u$		
K_2Cl_2	3.0	=	$6.0 \cdot u$		
moles H_2O	10^3	=	$10^3 \cdot u$	+	v

$u = 0.5, v = 500.$

The above solution 2 is saturated with respect to NaCl. Thus halite will be precipitated if the evaporation of water continues. Assuming that solution 3 is attained the quantity of the precipitated halite and the evaporated water is calculated from the equations

moles	solution 2		halite		solution 3		vapor
Na_2Cl_2	26.2	=	u	+	$49.5 \cdot v$		
K_2Cl_2	3.0	=			$12.0 \cdot v$		
H_2O	500	=			$10^3 \cdot v$	+	w

$u = 13.8, v = 0.25, w = 250.$

If solution 4 is attained upon further evaporation the quantity of the precipitated halite and the loss of water is

moles	solution 3		halite		solution 4		vapor
Na_2Cl_2	12.4	=	u	+	$46.6 \cdot v$		
K_2Cl_2	3.0	=			$18.2 \cdot v$		
H_2O	250	=			$10^3 \cdot v$	+	w

$u = 4.7, v = 0.1648, w = 85.2.$

From solution 4 both, halite and sylvite will precipitate upon further evaporation. At dryness there will be 52.4 mol halite and 6 mol sylvite, corresponding to 26.2 mol Na_2Cl_2 and 3 mol K_2Cl_2 of the initial solution.

II.6.3.3
Polythermal Precipitation of Sylvite

We consider the separation of sylvite from a halite-sylvite rock which has a ratio of $Na_2Cl_2/K_2Cl_2 = 57.1/5.0$. This may be achieved by dissolving a portion of the rock at a temperature of 90 °C (path 0-1 in Fig. 2.24). Separating the liquid from the solid, and evaporating solution 1, its composition changes along the path 1-2, precipitating halite. If point 2 is attained the solution is saturated with respect to halite and sylvite. After separating solution 2 from the previously precipitated halite and cooling it to 20 °C, it will change its composition along the path 2-3. As the path is located in the stability field of sylvite this solid will precipitate.

The compositions of the solutions are

solution	°C	$\{Na_2Cl_2\}$	$\{K_2Cl_2\}$
1	90	57.1	5.0
2	90	42.5	39.8
3	20	42.5	19.6

The quantity of halite precipitated along path 1-2 and the loss of water is

moles	90 °C solution 1		halite		90 °C solution 2		vapor
Na_2Cl_2	57.1	=	u	+	$42.5 \cdot v$		
K_2Cl_2	5.0	=			$39.8 \cdot v$		
H_2O	10^3	=			$10^3 \cdot v$	+	w

$u = 51.8, v = 0.1256, w = 874.4.$

The quantity of the precipitated sylvite is

moles	90 °C solution 2		sylvite		20 °C solution 3
Na_2Cl_2	5.34	=			$42.5 \cdot v$
K_2Cl_2	5.0	=	u	+	$19.6 \cdot v$
H_2O	125.6	=			$10^3 \cdot v$

$u = 2.54, v = 0.1256.$

It appears that about one half of the potassium chloride of the initial solution has been precipitated.

II.6.4
System 3.3 K_2Cl_2-$MgCl_2$-H_2O

bi: bischofite ($MgCl_2 \cdot 6H_2O$)
ca: carnallite ($KMgCl_3 \cdot 6H_2O$)
sy: sylvite (KCl)

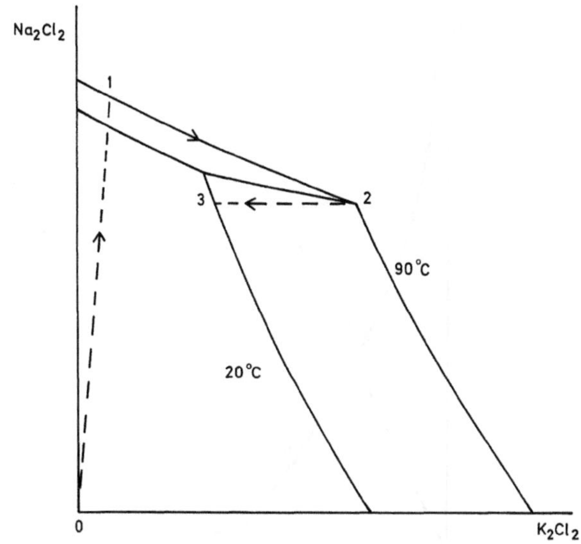

Fig. 2.24 System 3.1 Na_2Cl_2-K_2Cl_2-H_2O. Paths upon the polythermal precipitation of sylvite. 0-1: dissolution of a mixture of halite and sylvite. 1-2: precipitation of halite upon evaporation. 2-3: precipitation of sylvite upon cooling

II.6.4.1 Isothermal Evaporation

Figure 2.25 shows the 55 °C isotherm of the above system. Carnallite has a ratio $Mg^{2+}/K_2^{2+} = 2$ and is stable along the line 2-4. As any point on this line has a ratio $Mg^{2+}/K_2^{2+} > 2$, carnallite dissolves incongruently. The compositions of some of the saturated solutions are

solution	solids	$\{K_2Cl_2\}$	$\{MgCl_2\}$	C_T
1	bi	0	112.9	112.9
2	bi + ca	1.3	112.9	114.2
3	ca	3.8	95.0	98.8
4	ca + sy	8.0	78.2	86.2
5	sy	19.0	50.0	69.0
6	sy	37.4	20.0	57.4
7	sy	53.6	0	53.6

Initial Solution in the Carnallite Field

If solution 3 is evaporated its composition changes along the line 3-2 while carnallite is precipitated (Fig. 2.25). The quantities of carnallite and the vapor loss are calculated from the equations

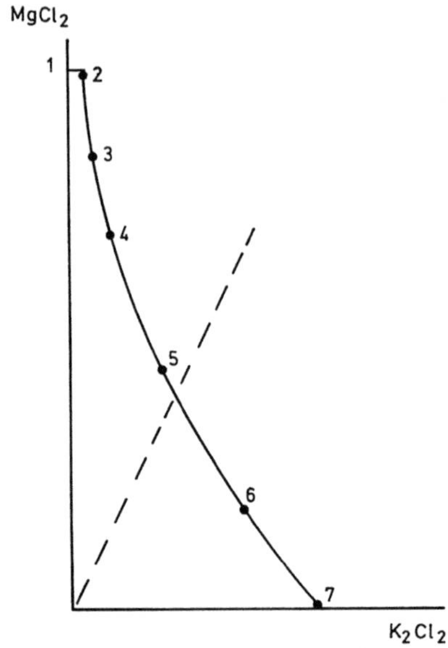

Fig. 2.25 System 3.3 K_2Cl_2-$MgCl_2$-H_2O. The slope of the broken line is $Mg^{2+}/K_2^{2+} = 2$ and corresponds to the cation composition of carnallite. 2: bischofite + carnallite. 4: carnallite + sylvite

moles	solution 3		carnallite		solution 2		vapor
K_2Cl_2	3.8	=	$0.5 \cdot u$	+	$1.3 \cdot v$		
$MgCl_2$	95.0	=	u	+	$112.9 \cdot v$		
H_2O	10^3	=	$6 \cdot u$	+	$10^3 \cdot v$	+	w

$u = 5.54, v = 0.7924, w = 174.4.$

It appears that 3.3% of the initial water is bound in the carnallite, 79.3% is left in the solution, and 17.4% has been evaporated. If solution 3 is evaporated to dryness, 7.6 mol carnallite + 87.4 mol bischofite will form. Carnallite and bischofite will bind 45.6 and 524.4 mol H_2O, respectively, and 430 mol H_2O will have evaporated.

It may be worthwhile to reconsider the above example with respect to the ion fractions (Eqs. II.3.1-II.3.3)

	f_{K_2}	f_{Mg}	C_T
solution 3	0.03846	0.96154	98.8
solution 2	0.01138	0.98862	114.2
carnallite	1/3	2/3	

Using the lever rule the equations

$$f_{K_2}{}^{sol.3} = f_{K_2}{}^{ca} \cdot X^{ca} + f_{K_2}{}^{sol.2} \cdot X^{sol.2}$$

$$f_{Mg}{}^{sol.3} = f_{Mg}{}^{ca} \cdot X^{ca} + f_{Mg}{}^{sol.2} \cdot X^{sol.2}$$

are obtained, where X^{ca} and $X^{sol.2}$ are the fractions of carnallite and solution 2, respectively. From the expressions

$$0.03846 = \frac{1}{3} X^{ca} + 0.01138 \cdot X^{sol.2}$$

$$0.96154 = \frac{2}{3} X^{ca} + 0.98862 \cdot X^{sol.2}$$

it is calculated that $X^{ca} = 0.08411$, and $X^{sol.2} = 0.91589$.

Multiplying the above equations by $C_T = 98.8$ yields the mass balances

moles	solution 3		carnallite		solution 2
K_2^{2+}	3.8	=	2.77	+	1.03
Mg^{2+}	95.0	=	5.54	+	89.46

Initial Solution in the Sylvite Field with $Mg^{2+}/K_2^{2+} > 2$

Solution 5 (Fig. 2.25) has a ratio $Mg^{2+}/K_2^{2+} > 2$ and is stable with sylvite. Upon the evaporation of water, sylvite will be formed along the line 5-4. At point 4 carnallite becomes stable. Upon further evaporation the previously precipitated sylvite will react via dissolution-precipitation to yield carnallite. During this reaction the composition of the solution stays constant. If sylvite is consumed the solution changes its composition along the line 4-2, precipitating carnallite.

Evaporating along the line 5-4 yields the quantities

moles	solution 5		sylvite		solution 4		vapor
K_2Cl_2	19.0	=	u	+	$8.0 \cdot v$		
$MgCl_2$	50.0	=			$78.2 \cdot v$		
H_2O	10^3	=			$10^3 \cdot v$	+	w

$u = 13.9, v = 0.6394, w = 360.6.$

The quantities upon evaporation along path 5-2 are

moles	solution 5		carnallite		solution 2		vapor
K_2Cl_2	19.0	=	$0.5 \cdot u$	+	$1.3 \cdot v$		
$MgCl_2$	50.0	=	u	+	$112.9 \cdot v$		
H_2O	10^3	=	$6 \cdot u$	+	$10^3 \cdot v$	+	w

$u = 37.72, v = 0.1088, w = 664.9.$

Initial Solution in the Sylvite Field with $Mg^{2+}/K_2^{2+} < 2$

Solution 6 (Fig. 2.25) is in equilibrium with sylvite, but has a ratio $Mg^{2+}/K_2^{2+} < 2$, or an excess of K_2Cl_2 with respect to the composition of carnallite. Thus, point 4 cannot be passed upon evaporation. The quantity of sylvite precipitated along path 6-4 is

moles	solution 6		sylvite		solution 4		vapor
K_2Cl_2	37.4	=	u	+	$8.0 \cdot v$		
$MgCl_2$	20.0	=			$78.2 \cdot v$		
H_2O	10^3	=			$10^3 \cdot v$	+	w

$u = 35.4, v = 0.2558, w = 744.2.$

Evaporating solution 6 to dryness yields 54.8 mol sylvite and 20 mol carnallite.

II.6.4.2
Dissolution of Carnallite

If an excess of carnallite is dissolved in water the composition of the solution evolves along the broken line in Fig. 2.25 at a ratio of $Mg^{2+}/K_2^{2+} = 2$. As the saturated solution is not stable with carnallite, sylvite will precipitate and carnallite will continue to dissolve until point 4 is reached. The quantities of carnallite, sylvite, and water are calculated from the equations

moles	solution 4		sylvite		carnallite		water
K_2Cl_2	8.0	=	u	+	$0.5 \cdot v$		
$MgCl_2$	78.2	=			v		
H_2O	10^3	=			$6 \cdot v$	+	w

$u = -31.1, v = 78.2, w = 530.8.$

In order to produce solution 4 from carnallite and water a quantity of 78.2 mol carnallite and 530.8 mol H_2O are required. The negative value $u = -31.1$ indicates that a quantity of 62.2 mol sylvite is precipitated.

The above example is re-considered with respect to ion fractions

	f_{K_2}	f_{Mg}	C_T
solution 4	0.09281	0.90719	86.2
carnallite	1/3	2/3	
sylvite	1		

The equations

$$f_{K_2}{}^{sol.4} = f_{K_2}{}^{ca} \cdot X^{ca} + f_{K_2}{}^{sy} \cdot X^{sy}$$

$$f_{Mg}{}^{sol.4} = f_{Mg}{}^{ca} \cdot X^{ca}$$

yield

$$X^{ca} = \frac{3}{2} 0.90719 = 1.36079$$

and

$$X^{sy} = 0.09281 - \frac{1.36079}{3} = -0.36079$$

Multiplying by C_T yields the mass balances

moles	solution 4		carnallite		sylvite
K_2^{2+}	8.0	=	39.1	–	31.1
Mg^{2+}	78.2	=	78.2		

II.6.5 System 4.1 Na_2Cl_2-K_2Cl_2-$MgCl_2$-H_2O

bi: bischofite ($MgCl_2·6H_2O$)
ca: carnallite ($KMgCl_3·6H_2O$)
ha: halite (NaCl)
sy: sylvite (KCl)

II.6.5.1 Calculation of Ion Fractions from Concentrations

At 20 °C the compositions of the solutions at the invariant points are

	$\{Na_2Cl_2\}$	$\{K_2Cl_2\}$	$\{MgCl_2\}$
ha + sy	46.6	18.2	0
bi + ha	0.6	0	102.6
bi + ca	0	0.8	102.3
ca + sy	0	5.2	70.9
bi + ca + ha	0.8	0.2	103.1
ca + ha + sy	4.1	5.5	69.7

The ion fractions with respect to a projection from the H_2O-point are (Eqs. II.4.1-II.4.4)

	f_{Na_2}	f_{K_2}	f_{Mg}	C_T
ha + sy	0.7191	0.2809	0	64.8
bi + ha	0.0058	0	0.9942	103.2
bi + ca	0	0.0078	0.9922	103.1
ca + sy	0	0.0683	0.9317	76.1
bi + ca + ha	0.0077	0.0019	0.9904	104.1
ca + ha + sy	0.0517	0.0694	0.8789	79.3

The relative compositions at saturation with respect to halite are (Eqs. II.4.6-II.4.9)

	F_{K_2}	F_{Mg}	C_P	F_{Na_2}
ha + sy	1	0	18.2	2.56
bi + ha	0	1	102.6	$5.58 \cdot 10^{-3}$
bi + ca	0.0078	0.9922	103.1	0
ca + sy	0.0683	0.9317	76.1	0
bi + ca + ha	0.0019	0.9981	103.3	$7.73 \cdot 10^{-3}$
ca + ha + sy	0.0731	0.9269	75.2	$5.45 \cdot 10^{-2}$

II.6.6
System 4.2 Na_2SO_4-K_2SO_4-$MgSO_4$-H_2O

ac: arcanite (K_2SO_4)
gs: glaserite ($K_3Na(SO_4)_2$)
pc: picromerite ($K_2Mg(SO_4)_2 \cdot 6H_2O$)

II.6.6.1
Isothermal Evaporation

In Fig. 2.26 solution 1 is supposed to be undersaturated. Once saturation is attained upon evaporation, arcanite will be precipitated along the path 1-2. At point 2 glaserite becomes stable, and the solution evolves along the path 2-3 precipitating arcanite + glaserite. The composition of the solutions and solids is given in terms of ion fractions (Eqs. II.4.1-II.4.4)

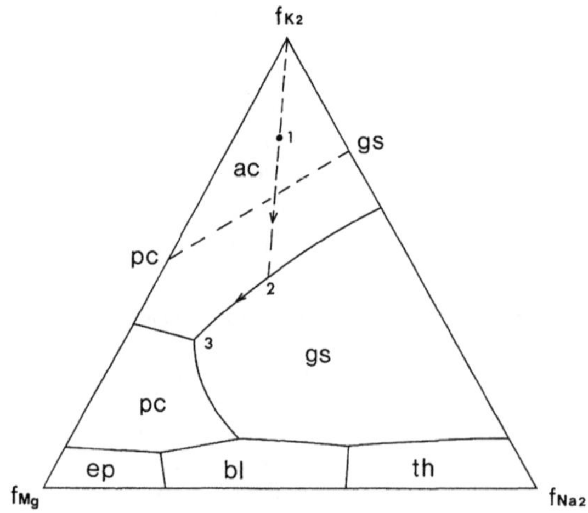

Fig. 2.26 System 4.2 Na_2SO_4-K_2SO_4-$MgSO_4$-H_2O. Projection of the 35 °C-isotherm from the H_2O-point. ac: arcanite, bl: bloedite, ep: epsomite, gs: glaserite, pc: picromerite, th: thenardite

35 °C	f_{Na_2}	f_{K_2}	f_{Mg}	C_T
solution 1	0.1067	0.7533	0.1400	15.0
solution 3	0.1409	0.3268	0.5323	51.1
arcanite		1		
glaserite	0.25	0.75		
picromerite		0.5	0.5	

The equations

$$f_{Na_2}{}^{sol.1} = f_{Na_2}{}^{gs} \cdot X^{gs} + f_{Na_2}{}^{sol.3} \cdot X^{sol.3}$$

$$f_{K_2}{}^{sol.1} = f_{K_2}{}^{gs} \cdot X^{gs} + f_{K_2}{}^{ac} \cdot X^{ac} + f_{K_2}{}^{sol.3} \cdot X^{sol.3}$$

$$f_{Mg}{}^{sol.1} = f_{Mg}{}^{sol.3} \cdot X^{sol.3}$$

yield the fractions $X^{sol.3} = 0.2630$, $X^{gs} = 0.2786$, and $X^{ac} = 0.4584$.

Multiplying the above equations by $C_T = 15.0$ yields the balances

moles	solution 1		arcanite		glaserite		solution 3
$Na_2{}^{2+}$	1.6	=			1.04	+	0.56
$K_2{}^{2+}$	11.3	=	6.9	+	3.12	+	1.28
Mg^{2+}	2.1	=					2.1

If solution 1 is evaporated to dryness arcanite + glaserite + picromerite will form. The equations

$$f_{Na_2}{}^{sol.1} = f_{Na_2}{}^{gs} \cdot X^{gs}$$

$$f_{K_2}{}^{sol.1} = f_{K_2}{}^{ac} \cdot X^{ac} + f_{K_2}{}^{gs} \cdot X^{gs} + f_{K_2}{}^{pc} \cdot X^{pc}$$

$$f_{Mg}{}^{sol.1} = f_{Mg}{}^{pc} \cdot X^{pc}$$

yield the fractions $X^{gs} = 0.4268$, $X^{pc} = 0.2800$, and $X^{ac} = 0.2932$.

II.6.6.2
Dissolution of Glaserite and Picromerite

Both glaserite and picromerite are dissolving incongruently (Fig. 2.26). If an excess of the two solids is dissolved in water arcanite will be precipitated and solution 3 will be produced. From the ion fractions

	f_{Na_2}	f_{K_2}	f_{Mg}	C_T
solution 3	0.1409	0.3268	0.5323	51.1
arcanite		1		
glaserite	0.25	0.75		
picromerite		0.5	0.5	

and the equations

$$f_{Na_2}{}^{sol.3} = f_{Na_2}{}^{gs} \cdot X^{gs}$$

$$f_{K_2}{}^{sol.3} = f_{K_2}{}^{ac} \cdot X^{ac} + f_{K_2}{}^{gs} \cdot X^{gs} + f_{K_2}{}^{pc} \cdot X^{pc}$$

$$f_{Mg}{}^{sol.3} = f_{Mg}{}^{pc} \cdot X^{pc}$$

it is calculated that

$$X^{pc} = 1.0646, X^{gs} = 0.5636, \text{and } X^{ac} = -0.6282.$$

The negative sign of X^{ac} indicates that arcanite is produced by the reaction of water with glaserite and picromerite.
Multiplying the above equations by $C_T = 51.1$ yields

moles	solution 3		ac		gs		pc
$Na_2{}^{2+}$	7.2	=			7.2		
$K_2{}^{2+}$	16.7	=	–32.1	+	21.6	+	27.2
Mg^{2+}	27.2	=					27.2

It is assumed that in the above example the compositions are given in terms of a projection from the K_2SO_4 point onto the line f_{Mg}-f_{Na_2} (saturation with respect to arcanite). The relative compositions are (Eqs. II.4.6-II.4.9)

	F_{Na_2}	F_{Mg}	C_P	F_{K_2}
solution 3	0.2093	0.7907	34.4	0.4855
glaserite	1			3
picromerite		1		1

The quantities of neutral salts in the solids are

glaserite
moles: $Na_2SO_4 = F_{Na_2}{}^{sol.3} \cdot C_P = 7.2$

moles: $K_2SO_4 = 3 \cdot F_{Na_2}{}^{sol.3} \cdot C_P = 21.6$

picromerite
moles: $MgSO_4 = K_2SO_4 = F_{Mg}{}^{sol.3} \cdot C_P = 27.2$

arcanite
moles: $K_2SO_4 = (F_{K_2}{}^{sol.3} - 3 \cdot F_{Na_2}{}^{sol.3} - F_{Mg}{}^{sol.3}) \cdot C_P = -32.1.$

II.6.7
System 4.3 Na_2Cl_2-K_2Cl_2-Na_2SO_4-K_2SO_4-H_2O

ac: arcanite (K_2SO_4)
gs: glaserite ($K_3Na(SO_4)_2$)
ha: halite (NaCl)
sy: sylvite (KCl)
th: thenardite (Na_2SO_4)

II.6.7.1
Calculation of Ion Fractions from Concentrations

Figure 2.27 shows a projection of the system at 50 °C. The compositions of the invariant solutions are

solution	solids	$\{Na_2Cl_2\}$	$\{K_2Cl_2\}$	$\{Na_2SO_4\}$
1	ac + gs + sy	13.2	43.5	3.1
2	gs + ha + sy	41.6	27.6	4.0
3	gs + ha + th	44.5	14.2	12.2

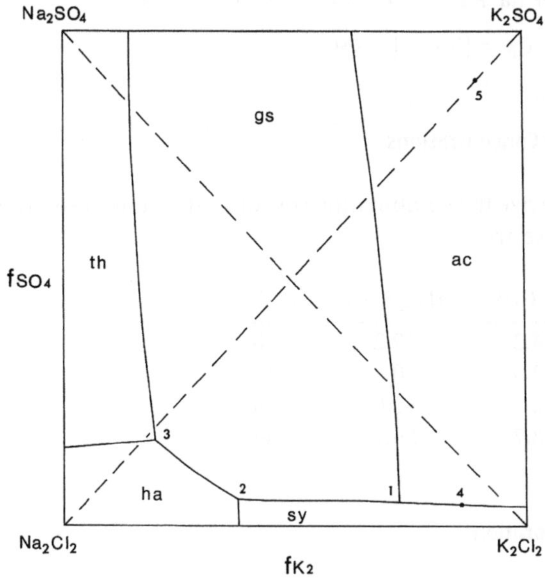

Fig. 2.27 System 4.3 Na_2Cl_2-K_2Cl_2-Na_2SO_4-K_2SO_4-H_2O. Projection of the 50 °C-isotherm from the H_2O-point. ac: arcanite, gs: glaserite, ha: halite, sy: sylvite, th: thenardite

The ion fractions with respect to a projection from the H_2O-point are (Eqs. II.4.10, II.4.15-II.4.18)

solution	solids	f_{Na_2}	f_{SO_4}	C_T
1	ac + gs + sy	0.2726	0.0518	59.8
2	gs + ha + sy	0.6230	0.0546	73.2
3	gs + ha + th	0.7997	0.1721	70.9

$f_{K_2} = 1 - f_{Na_2}, f_{Cl_2} = 1 - f_{SO_4}$.

The relative compositions at saturation with respect to halite are (Eqs. II.4.20-II.4.25)

solution	solids	F_{K_2}	C_P	F_{Na_2}
2	gs + ha + sy	0.8734	31.6	1.4430
3	gs + ha + th	0.5379	26.4	2.1477

$F_{SO_4} = 1 - F_{K_2}, C_P = \{K_2^{2+}\} + \{SO_4^{2-}\}$.

The relative compositions at saturation with respect to sylvite are

solution	solids	F_{Na_2}	C_P	F_{K_2}
1	ac + gs + sy	0.8402	19.4	2.2423
2	gs + ha + sy	0.9194	49.6	0.5565

$F_{SO4} = 1 - F_{Na2}, C_P = \{Na_2^{2+}\} + \{SO_4^{2-}\}$.

II.6.7.2
Combinations of Concentrations

We now summarize the combinations of concentrations of solution 3. According to Table 2.2 these are

$\{Na_2Cl_2\}$	$\{K_2Cl_2\}$	$\{Na_2SO_4\}$	$\{K_2SO_4\}$
44.5	14.2	12.2	0
56.7	2.0	0	12.2
58.7	0	−2.0	14.2
0	58.7	56.7	−44.5

II.6.7.3
Isothermal Evaporation

Solution 4 (Fig. 2.27) is supposed to be saturated. If it is evaporated so that solution 1 is attained arcanite and sylvite will be precipitated. The composition of the solutions are

	$\{Na_2Cl_2\}$	$\{K_2Cl_2\}$	$\{Na_2SO_4\}$
solution 4	5.4	48.0	2.5
solution 1	13.2	43.5	3.1

The quantities of the solids precipitated along the path 4-1 and the evaporated water are calculated from the equations

	solution 4		arcanite		sylvite		solution 1		vapor
Na_2^{2+}	7.9	=					$16.3 \cdot w$		
K_2^{2+}	48.0	=	u	+	v	+	$43.5 \cdot w$		
Cl_2^{2-}	53.4	=			v	+	$56.7 \cdot w$		
SO_4^{2-}	2.5	=	u			+	$3.1 \cdot w$		
H_2O	10^3	=					$10^3 \cdot w$	+	x

$u = 1.0, v = 25.92, w = 0.4847, x = 515.3.$

The composition of solution 1 is (moles)

Na_2Cl_2	K_2Cl_2	Na_2SO_4	H_2O
6.40	21.08	1.50	484.7

It is assumed that solution 4 is evaporated to yield solution 2. At point 1 arcanite which was precipitated along the path 4-1 will react to glaserite, and along path 1-2 glaserite + sylvite will be precipitated. The compositions of the solutions and solids are given in terms of a projection from the H_2O point

	f_{Na_2}	f_{K_2}	f_{Cl_2}	f_{SO_4}	C_T
solution 4	0.1413	0.8587	0.9553	0.0447	55.9
solution 2	0.6230	0.3770	0.9454	0.0546	73.2
sylvite		1	1		
glaserite	0.25	0.75		1	

From the equations

$$f_{Na_2}^{sol.4} = f_{Na_2}^{gs} \cdot X^{gs} + f_{Na_2}^{sol.2} \cdot X^{sol.2}$$

$$f_{K_2}^{sol.4} = X^{sy} + f_{K_2}^{gs} \cdot X^{gs} + f_{K_2}^{sol.2} \cdot X^{sol.2}$$

$$f_{Cl_2}^{sol.4} = X^{sy} + f_{Cl_2}^{sol.2} \cdot X^{sol.2}$$

$$f_{SO_4}^{sol.4} = f_{SO_4}^{gs} \cdot X^{gs} + f_{SO_4}^{sol.2} \cdot X^{sol.2}$$

or

$$0.1413 = 0.25 \cdot X^{gs} + 0.6230 \cdot X^{sol.2}$$

$$0.8587 = X^{sy} + 0.75 \cdot X^{gs} + 0.3770 \cdot X^{sol.2}$$

$$0.9553 = X^{sy} + 0.9454 \cdot X^{sol.2}$$

$$0.0447 = X^{gs} + 0.0546 \cdot X^{sol.2}$$

it is calculated that $X^{sol.2} = 0.2136$, $X^{sy} = 0.7534$, and $X^{gs} = 0.0330$. Multiplying the above equations by $C_T = 55.9$ yields the balances

moles	solution 4		sylvite		glaserite		solution 2
Na_2^{2+}	7.9	=			0.46	+	7.44
K_2^{2+}	48.0	=	42.12	+	1.39	+	4.50
Cl_2^{2-}	53.4	=	42.12	+		+	11.29
SO_4^{2-}	2.5	=			1.85	+	0.65

Evaporating solution 4 to dryness yields a mixture of glaserite, halite, and sylvite. The composition of the solution and the solids is given in terms of a projection from the K_2Cl_2-point, e. g. at saturation with respect to sylvite

	F_{Na_2}	F_{SO_4}	C_P	F_{K_2}
solution 4	0.7596	0.2404	10.4	4.6154
halite	1			
glaserite	0.2	0.8		0.6

The fractions of halite and glaserite are calculated from the equations

$$F_{Na_2}{}^{sol.4} = F_{Na_2}{}^{ha} \cdot X^{ha} + F_{Na_2}{}^{gs} \cdot X^{gs} = 0.7596 = X^{ha} + 0.2 \cdot X^{gs}$$

$$F_{SO_4}{}^{sol.4} = F_{SO_4}{}^{gs} \cdot X^{gs} = 0.2404 = 0.8 \cdot X^{gs}$$

It appears that $X^{gs} = 0.3005$, and $X^{ha} = 0.6995$.
From these quantities it is calculated that

$$\text{moles: } Na_2^{2+(ha)} = Cl_2^{2-(ha)} = X^{ha} \cdot C_P = 7.275$$

$$\text{moles: } Na_2^{2+(gs)} = F_{Na_2}{}^{gs} \cdot X^{gs} \cdot C_P = 0.625$$

$$\text{moles: } SO_4^{2-(gs)} = F_{SO_4}{}^{gs} \cdot X^{gs} \cdot C_P = 2.5$$

The quantity of K_2^{2+} in sylvite (moles) is obtained from the balance

$$K_2^{2+(sy)} = Cl_2^{2-(sy)} = (F_{K_2}{}^{sol.4} - F_{K_2}{}^{gs} \cdot X^{gs}) \cdot C_P = 46.125.$$

Summarizing the results yields the balances

moles	solution 4		halite		sylvite		glaserite
Na_2^{2+}	7.9	=	7.275	+		+	0.625
K_2^{2+}	48.0	=			46.125	+	1.875
Cl_2^{2-}	53.4	=	7.275	+	46.125		
SO_4^{2-}	2.5	=					2.5

Solution 5 (Fig. 2.27) is supposed to be undersaturated. Evaporation to dryness yields a mixture of arcanite, glaserite, and sylvite. The composition of the

solution is given in terms of a projection from the K_2SO_4 point, e.g. at saturation with respect to arcanite

	$F_{Na_2} = F_{Cl_2}$	C_P	F_{K_2}
solution 5	0.5	3.0	4.5

The equations

$$\{Na_2^{2+}\} = \{Cl_2^{2-}\} = F_{Na_2} \cdot C_P = 1.5$$

$$\{K_2^{2+}\} = \{SO_4^{2-}\} = F_{K_2} \cdot C_P = 13.5$$

yield the balances

	solution 5		sylvite		arcanite		glaserite
Na_2^{2+}	1.5	=					$0.5 \cdot w$
K_2^{2+}	13.5	=	u	+	v	+	$1.5 \cdot w$
Cl_2^{2-}	1.5	=	u				
SO_4^{2-}	13.5	=			v	+	$2.0 \cdot w$

$u = 1.5, v = 7.5, w = 3.0.$

II.6.8
System 4.4 Na_2Cl_2-$MgCl_2$-Na_2SO_4-$MgSO_4$-H_2O

bl: bloedite ($Na_2Mg(SO_4)_2 \cdot 4H_2O$)
ep: epsomite ($MgSO_4 \cdot 7H_2O$)
ha: halite (NaCl)
mi: mirabilite ($Na_2SO_4 \cdot 10H_2O$)

II.6.8.1
Isothermal Evaporation

At 25 °C solution 1 is saturated with respect to bloedite and halite (Fig. 2.28). Upon evaporation solution 2 will be attained. The compositions of the solutions are

solution	solid	$\{Na_2Cl_2\}$	$\{MgCl_2\}$	$\{Na_2SO_4\}$
1	bl + ha	10.8	40.0	15.1
2	bl + ep + ha	−4.5	60.8	16.3

Along the path 1-2 bloedite and halite will be precipitated. The quantities of the solids, the evaporated water, and the ions remaining in solution are calculated from the equations

	solution 1		halite		bloedite		solution 2		vapor
Na_2^{2+}	25.9	=	u	+	v	+	$11.8 \cdot w$		
Mg^{2+}	40.0	=			v	+	$60.8 \cdot w$		
Cl_2^{2-}	50.8	=	u			+	$56.3 \cdot w$		
SO_4^{2-}	15.1	=			$2 \cdot v$	+	$16.3 \cdot w$		
H_2O	10^3	=			$4 \cdot v$	+	$10^3 \cdot w$	+	x

$u = 16.1, v = 2.53, w = 0.6163, x = 373.58.$

The solution contains 616.3 mol water, and 10.1 mol water are bound in bloedite.

It is assumed that the compositions of the above solutions are given in terms of a projection from the H_2O point (Eqs. II.4.10, II.4.15-II.4.18)

	f_{Na_2}	f_{Mg}	f_{Cl_2}	f_{SO_4}	C_T
solution 1	0.3930	0.6070	0.7709	0.2291	65.9
solution 2	0.1625	0.8375	0.7755	0.2245	72.6
halite	1		1		
bloedite	0.5	0.5		1	

The fractions of the solids and the solution are calculated from the equations

$$f_{Na_2}{}^{sol.1} = f_{Na_2}{}^{ha} \cdot X^{ha} + f_{Na_2}{}^{bl} \cdot X^{bl} + f_{Na_2}{}^{sol.2} \cdot X^{sol.2}$$

$$f_{Mg}{}^{sol.1} = f_{Mg}{}^{bl} \cdot X^{bl} + f_{Mg}{}^{sol.2} \cdot X^{sol.2}$$

$$f_{Cl_2}{}^{sol.1} = f_{Cl_2}{}^{ha} \cdot X^{ha} + f_{Cl_2}{}^{sol.2} \cdot X^{sol.2}$$

$$f_{SO_4}{}^{sol.1} = f_{SO_4}{}^{bl} \cdot X^{bl} + f_{SO_4}{}^{sol.2} \cdot X^{sol.2}$$

or

$$0.3930 = X^{ha} + 0.5 \cdot X^{bl} + 0.1625 \cdot X^{sol.2}$$

$$0.6070 = 0.5 \cdot X^{bl} + 0.8375 \cdot X^{sol.2}$$

$$0.7709 = X^{ha} + 0.7755 \cdot X^{sol.2}$$

$$0.2291 = X^{bl} + 0.2245 \cdot X^{sol.2}$$

which yield $X^{sol.2} = 0.6790$, $X^{bl} = 0.0767$, and $X^{ha} = 0.2443$.

Multiplying the above equations by $C_T = 65.9$ yields the balances

moles	solution 1		halite		bloedite		solution 2
Na_2^{2+}	25.9	=	16.10	+	2.52	+	7.28
Mg^{2+}	40.0	=			2.52	+	37.48
Cl_2^{2-}	50.8	=	16.10			+	34.70
SO_4^{2-}	15.1	=			5.04	+	10.06

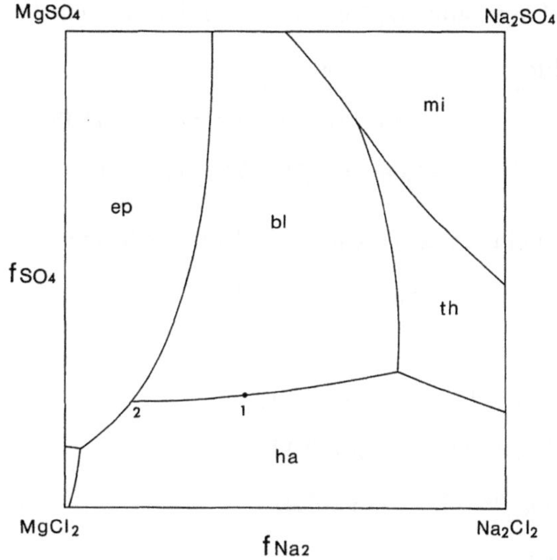

Fig. 2.28 System 4.4 Na_2Cl_2-$MgCl_2$-Na_2SO_4-$MgSO_4$-H_2O. Projection of the 25 °C-isotherm from the H_2O-point. bl: bloedite, ep: epsomite, ha: halite, mi: mirabilite, th: thenardite; lower left corner: other Mg-sulfates and bischofite

It is assumed that the compositions of solutions 1 and 2 are given with respect to a projection from the Na_2Cl_2 point (Eqs. II.4.20-II.4.25)

	F_{Mg}	F_{SO_4}	C_P	F_{Na_2}
solution 1	0.7260	0.2740	55.1	0.4701
solution 2	0.7886	0.2114	77.1	0.1530
bloedite	1/3	2/3		1/3

The fractions of the precipitated bloedite and solution 2 are obtained from the equations

$$F_{Mg}{}^{sol.1} = F_{Mg}{}^{bl} \cdot X^{bl} + F_{Mg}{}^{sol.2} \cdot X^{sol.2}$$

$$F_{SO_4}{}^{sol.1} = F_{SO_4}{}^{bl} \cdot X^{bl} + F_{SO_4}{}^{sol.2} \cdot X^{sol.2}$$

or

$$0.7260 = \frac{1}{3} X^{bl} + 0.7886 \cdot X^{sol.2}$$

$$0.2740 = \frac{2}{3} X^{bl} + 0.2114 \cdot X^{sol.2}$$

which yield $X^{bl} = 0.1375$, and $X^{sol.2} = 0.8625$.

The quantities of Mg^{2+} and SO_4^{2-} in bloedite and in solution 2 are

moles	bloedite	solution 2
Mg^{2+}	$F_{Mg}{}^{bl} \cdot X^{bl} \cdot C_P = 2.53$	$F_{Mg}{}^{sol.2} \cdot X^{sol.2} \cdot C_P = 37.48$
SO_4^{2-}	$F_{SO_4}{}^{bl} \cdot X^{bl} \cdot C_P = 5.06$	$F_{SO_4}{}^{sol.2} \cdot X^{sol.2} \cdot C_P = 10.05$

The quantity of halite (moles) is calculated from the balance

$$Na_2{}^{2+(sol.1)} = Na_2{}^{2+(ha)} + Na_2{}^{2+(bl)} + Na_2{}^{2+(sol.2)}$$

As

$$Na_2{}^{2+(sol.1)} = F_{Na_2}{}^{sol.1} \cdot C_P = 25.9$$

$$Na_2{}^{2+(bl)} = F_{Na_2}{}^{bl} \cdot X^{bl} \cdot C_P = 2.53$$

$$Na_2{}^{2+(sol.2)} = F_{Na_2}{}^{sol.2} \cdot X^{sol.2} \cdot C_P = 7.27$$

it is calculated that

$$Na_2{}^{2+(ha)} = Cl_2{}^{2-(ha)} = 25.9–2.53–7.27 = 16.1.$$

II 6.8.2
Generation of a Solution from Hydrated Solids

At 6.2 °C the stable solids are bloedite, epsomite, mirabilite, and halite. As bloedite is less hydrated than epsomite and mirabilite, a solution saturated with respect to the above four solids will be formed if a dry mixture of epsomite + mirabilite + halite is heated to a temperature of 6.2 °C. The composition of the solution is

$\{Na_2Cl_2\}$	$\{MgCl_2\}$	$\{MgSO_4\}$
30.2	16.3	18.1

The quantities of the solids are obtained from the equations

	solution		halite		mirabilite		epsomite		bloedite
$Na_2{}^{2+}$	30.2	=	u	+	v			+	x
Mg^{2+}	34.4	=					w	+	x
$Cl_2{}^{2-}$	46.5	=	u						
$SO_4{}^{2-}$	18.1	=			v	+	w	+	$2 \cdot x$
H_2O	10^3	=			$10 \cdot v$	+	$7 \cdot w$	+	$4 \cdot x$

u = 46.5, v = 54.6, w = 105.3, x = –70.9.

It appears that 93.0 mol halite + 54.6 mol mirabilite + 105.3 mol epsomite will react to yield 70.9 mol bloedite and a saturated solution.

II.6.9 System 5 Na_2Cl_2-K_2Cl_2-$MgCl_2$-Na_2SO_4-K_2SO_4-$MgSO_4$-H_2O

bi: bischofite ($MgCl_2 \cdot 6H_2O$), bl: bloedite ($Na_2Mg(SO_4)_2 \cdot 4H_2O$)
ca: carnallite ($KMgCl_3 \cdot 6H_2O$), ep: epsomite ($MgSO_4 \cdot 7H_2O$)
gs: glaserite ($K_3Na(SO_4)_2$), ha: halite (NaCl)
hx: hexahydrite ($MgSO_4 \cdot 6H_2O$), ka: kainite ($KMgClSO_4 \cdot 11/4\ H_2O$)
ks: kieserite ($MgSO_4 \cdot H_2O$), le: leonite ($K_2Mg(SO_4)_2 \cdot 4H_2O$)
lg: langbeinite ($K_2Mg_2(SO_4)_3$), sy: sylvite (KCl)
pc: picromerite ($K_2Mg(SO_4)_2 \cdot 6H_2O$), th: thenardite ($Na_2SO_4$)

II.6.9.1 Calculation of Ion Fractions from Concentrations

At 25 °C solution R is in equilibrium with carnallite, kainite, kieserite, and halite (Fig. 2.29). The composition of the solution is

$\{Na_2Cl_2\}$	$\{K_2Cl_2\}$	$\{MgCl_2\}$	$\{MgSO_4\}$
2.4	1.4	80.6	6.2

The combinations of concentrations are (Tables 2.4 and 2.5)

$\{Na_2Cl_2\}$	$\{K_2Cl_2\}$	$\{MgCl_2\}$	$\{Na_2SO_4\}$	$\{K_2SO_4\}$	$\{MgSO_4\}$
0	0	84.4	2.4	1.4	2.4
0	84.4	0	2.4	−83.0	86.8
0	1.4	83.0	2.4	0	3.8
0	−2.4	86.8	2.4	3.8	0
84.4	0	0	−82.0	1.4	86.8
2.4	0	82.0	0	1.4	4.8
−2.4	0	86.8	4.8	1.4	0
2.4	82.0	0	0	−80.6	86.8
83.0	1.4	0	−80.6	0	86.8
2.4	1.4	80.6	0	0	6.2
2.4	−4.8	86.8	0	6.2	0
−3.8	1.4	86.8	6.2	0	0

The total concentration, the water value, and the ion concentrations are (Eqs. II.5.1-II.5.7)

$$C_T = \{Na_2^{2+}\} + \{K_2^{2+}\} + \{Mg^{2+}\} = \{Cl_2^{2-}\} + \{SO_4^{2-}\} = 90.6$$

$$W_T = \frac{10^5}{C_T} = 1103.8$$

$\{Na_2^{2+}\} = \{Na_2Cl_2\} = 2.4$
$\{K_2^{2+}\} = \{K_2Cl_2\} = 1.4$
$\{Mg^{2+}\} = \{MgCl_2\} + \{MgSO_4\} = 86.8$
$\{Cl_2^{2-}\} = \{Na_2Cl_2\} + \{K_2Cl_2\} + \{MgCl_2\} = 84.4$
$\{SO_4^{2-}\} = \{MgSO_4\} = 6.2$

The relative composition of the solution and the solids with respect to Fig. 2.19 is (Eqs. II.5.1, II.5.8-II.5.12)

	f_{Na_2}	f_{K_2}	f_{Mg}	f_{Cl_2}	f_{SO_4}
solution R	0.0265	0.0155	0.9580	0.9316	0.0684
carnallite		1/3	2/3	1	
kainite		1/3	2/3	1/3	2/3
kieserite			1		1
halite	1			1	

The ion concentration, the water value, and the ion fractions with respect to a projection from the Na_2Cl_2 point onto the triangle F_{SO_4}-F_{K_2}-F_{Mg} are (Eqs. II.5.13-II.5.19)

$C_P = \{K_2^{2+}\} + \{Mg^{2+}\} + \{SO_4^{2-}\} = 94.4, W_P = 1059.3$

	F_{K_2}	F_{Mg}	F_{SO_4}	F_{Na_2}
solution R	0.0148	0.9195	0.0657	0.0254
carnallite	1/3	2/3		
kainite	1/5	2/5	2/5	
kieserite		1/2	1/2	

II.6.9.2 Crystallization and Transition Lines

Figure 2.29 shows a projection of the 25 °C isotherm of the system at saturation with respect to halite. The lines between any two points are either crystallization or transition lines. In the first case the salt points are located on opposite sides of a phase boundary or its extension (e. g. the glaserite-sylvite or the kainite-sylvite boundary). In the second case the salt points are located on the same side of a phase boundary (e. g. the bloedite-epsomite or the leonite-kainite boundary). It may be noted that congruent solubility occurs only with respect to the assemblage bischofite + carnallite + kieserite (point Z).

Thus, upon evaporation of solution 1 (lower diagram) sylvite + halite will be precipitated. If the composition of the solution attains point 2, kainite will become stable and will be precipitated together with sylvite and halite along the

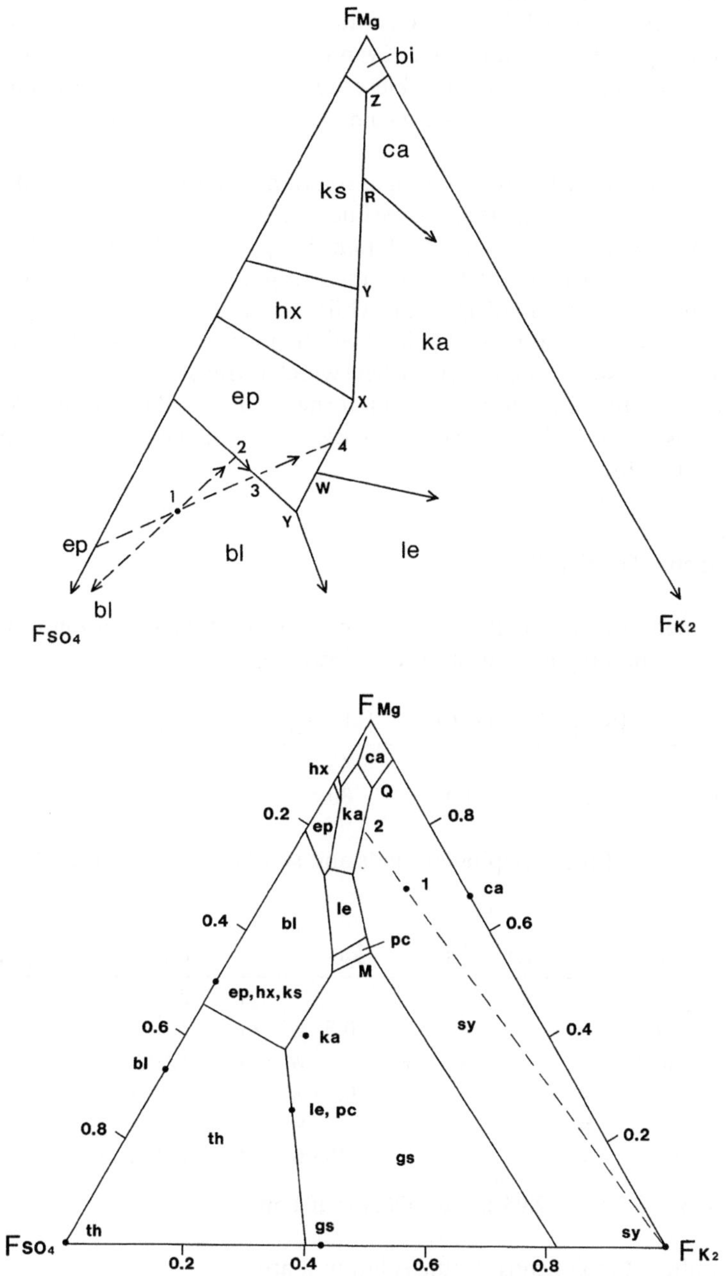

Fig. 2.29 System 5 Na_2Cl_2-K_2Cl_2-$MgCl_2$-Na_2SO_4-K_2SO_4-$MgSO_4$-H_2O. 25 °C-isotherm at saturation with respect to halite. bi: bischofite, bl: bloedite, ca: carnallite, ep: epsomite, gs: glaserite, hx: hexahydrite, ka: kainite, ks: kieserite, le: leonite, pc: picromerite, sy: sylvite, th: thenardite

line 2-Q. At point Q sylvite will be consumed, and kainite + carnallite + halite will be formed along Q-R. At point R kainite disappears and carnallite + kieserite + halite are precipitated along R-Z. Evaporating to dryness leads to the final assemblage bischofite + carnallite + kieserite + halite at point Z. This point where the solution evaporates at a constant composition is also denoted as the eutonic point.

On the other hand, the evaporation of solution 1 (upper diagram) leads to the formation of bloedite + halite along the line 1-2. At point 2 bloedite begins to disappear, and the composition of the solution changes along the line 2-3. At point 3 bloedite is consumed completely and the composition of the solution passes through the stability field of epsomite while epsomite + halite are precipitated. Meeting the line W-X at point 4 kainite will be formed additionally. At point X epsomite is transformed into hexahydrite which is transformed into kieserite at point Y. Along the line Y-R kainite + kieserite + halite will be formed. At point R kainite is consumed and carnallite + kieserite + halite will be precipitated along the line R-Z.

II.6.9.3
Evaporation of Solution W

It is assumed that solution R is attained upon evaporation of solution W at 25 °C (Fig. 2.29). The compositions of the solutions are

	$\{Na_2Cl_2\}$	$\{K_2Cl_2\}$	$\{MgCl_2\}$	$\{MgSO_4\}$
solution W	10.6	6.8	45.1	18.5
solution R	2.4	1.4	80.6	6.2

The quantities of the precipitated solids and solution R are calculated from the equations

moles	sol. W		ha	ks		ka		sol. R		vapor
Na_2^{2+}	10.6	=	u				+	$2.4 \cdot x$		
K_2^{2+}	6.8	=				$0.5 \cdot w$	+	$1.4 \cdot x$		
Mg^{2+}	63.6	=		v	+	w	+	$86.8 \cdot x$		
Cl_2^{2-}	62.5	=	u		+	$0.5 \cdot w$	+	$84.4 \cdot x$		
SO_4^{2-}	18.5	=		v	+	w	+	$6.2 \cdot x$		
H_2O	10^3	=		v	+	$2.75 \cdot w$	+	$10^3 \cdot x$	+	y

u = 9.26, v = 3.0, w = 12.03, x = 0.5596, y = 404.36.

The number of moles constituting solution R are

Na_2^{2+}	K_2^{2+}	Mg^{2+}	Cl_2^{2-}	SO_4^{2-}	H_2O
1.34	0.78	48.57	47.22	3.47	559.6

or

Na_2Cl_2	K_2Cl_2	$MgCl_2$	$MgSO_4$	H_2O
1.34	0.78	45.10	3.47	559.6

It is assumed that point Z is attained upon evaporation of solution W. The compositions of the solutions and solids are given in terms of a projection from the Na_2Cl_2 point

	F_{K_2}	F_{Mg}	F_{SO_4}	C_P	F_{Na_2}
solution W	0.0765	0.7154	0.2081	88.9	0.1192
solution Z	0.0019	0.9878	0.0103	106.6	0.0066
carnallite	1/3	2/3			
kieserite		1/2	1/2		

From the equations

$$F_{K_2}{}^{W} = F_{K_2}{}^{ca} \cdot X^{ca} + F_{K_2}{}^{Z} \cdot X^{Z}$$

$$F_{Mg}{}^{W} = F_{Mg}{}^{ca} \cdot X^{ca} + F_{Mg}{}^{ks} \cdot X^{ks} + F_{Mg}{}^{Z} \cdot X^{Z}$$

$$F_{SO_4}{}^{W} = F_{SO_4}{}^{ks} \cdot X^{ks} + F_{SO_4}{}^{Z} \cdot X^{Z}$$

or

$$0.0765 = \frac{1}{3} X^{ca} + 0.0019 \cdot X^{Z}$$

$$0.7154 = \frac{2}{3} X^{ca} + \frac{1}{2} X^{ks} + 0.9878 \cdot X^{Z}$$

$$0.2081 = \frac{1}{2} X^{ks} + 0.0103 \cdot X^{Z}$$

it is calculated that $X^{Z} = 0.3639$, $X^{ca} = 0.2274$, and $X^{ks} = 0.4087$.

From the concentration with respect to the projection $C_P = 88.9$ it is calculated that

moles	carnallite	kieserite	solution Z
$K_2{}^{2+}$	$F_{K_2}{}^{ca} \cdot X^{ca} \cdot C_P$		$F_{K_2}{}^{Z} \cdot X^{Z} \cdot C_P$
Mg^{2+}	$F_{Mg}{}^{ca} \cdot X^{ca} \cdot C_P$	$F_{Mg}{}^{ks} \cdot X^{ks} \cdot C_P$	$F_{Mg}{}^{Z} \cdot X^{Z} \cdot C_P$
$SO_4{}^{2-}$		$F_{SO_4}{}^{ks} \cdot X^{ks} \cdot C_P$	$F_{SO_4}{}^{Z} \cdot X^{Z} \cdot C_P$

or

moles	carnallite	kieserite	solution Z
$K_2{}^{2+}$	6.7		0.06
Mg^{2+}	13.4	18.17	31.95
$SO_4{}^{2-}$		18.17	0.33

The quantity of the precipitated halite is obtained from the balance

moles: $Na_2^{2+(W)} = Na_2^{2+(ha)} + Na_2^{2+(Z)}$

where

$$Na_2^{2+(W)} = F_{Na_2}{}^{W} \cdot C_P = 10.6$$

and

$$Na_2^{2+(Z)} = F_{Na_2}{}^{Z} \cdot X^{Z} \cdot C_P = 0.21$$

so that

$$Na_2^{2+(ha)} = 10.39.$$

As solution W is located in the triangle given by the salt points of bischofite, carnallite, and kieserite, evaporation to dryness leads to the formation of the above solids. Their quantities are obtained from the equations

$$F_{K_2}{}^{W} = F_{K_2}{}^{ca} \cdot X^{ca}$$

$$F_{Mg}{}^{W} = F_{Mg}{}^{ca} \cdot X^{ca} + F_{Mg}{}^{ks} \cdot X^{ks} + F_{Mg}{}^{bi} \cdot X^{bi}$$

$$F_{SO_4}{}^{W} = F_{SO_4}{}^{ks} \cdot X^{ks}$$

or

$$0.0765 = \frac{1}{3} X^{ca}$$

$$0.7154 = \frac{2}{3} X^{ca} + \frac{1}{2} X^{ks} + X^{bi}$$

$$0.2081 = \frac{1}{2} X^{ks}$$

It is calculated that $X^{ca} = 0.2295$, $X^{ks} = 0.4162$, and $X^{bi} = 0.3543$.

From the concentration with respect to the projection $C_P = 88.9$ it is calculated that

moles	carnallite	kieserite	bischofite
K_2^{2+}	6.8		
Mg^{2+}	13.6	18.5	31.5
SO_4^{2-}		18.5	

The quantity of the precipitated halite is

moles: $Na_2^{2+(ha)} = F_{Na_2}{}^{W} \cdot C_P = 10.6.$

II.6.9.4
Evaporation of Solution M

Solution M is located in the triangle given by the salt points of carnallite, kainite, and kieserite (Fig. 2.29). Thus, point R will be the endpoint of crystallization.

Evaporating solution M at 25 °C to point Q yields halite + sylvite + kainite. The compositions of the solutions are

	$\{Na_2Cl_2\}$	$\{K_2Cl_2\}$	$\{MgCl_2\}$	$\{MgSO_4\}$
solution M	24.3	14.5	21.3	13.9
solution Q	4.5	5.7	68.1	5.0

The quantities of the precipitated solids and the remaining solution are calculated from the equations

moles	sol. M		ha		sy		ka		sol. Q		vapor
Na_2^{2+}	24.3	=	u					+	$4.5 \cdot x$		
K_2^{2+}	14.5	=			v	+	$0.5 \cdot w$	+	$5.7 \cdot x$		
Mg^{2+}	35.2	=					w	+	$73.1 \cdot x$		
Cl_2^{2-}	60.1	=	u	+	v	+	$0.5 \cdot w$	+	$78.3 \cdot x$		
SO_4^{2-}	13.9	=					w	+	$5.0 \cdot x$		
H_2O	10^3	=					$2.75 \cdot w$	+	$10^3 \cdot x$	+	y

$u = 22.89, v = 6.55, w = 12.34, x = 0.3128, y = 653.3.$

It is assumed that point R is attained upon evaporation of solution M. The compositions of the solutions are given in terms of a projection from the Na_2Cl_2 point

	F_{K_2}	F_{Mg}	F_{SO_4}	C_P	F_{Na_2}
solution M	0.2280	0.5535	0.2185	63.6	0.3821
solution R	0.0148	0.9195	0.0657	94.4	0.0254
carnallite	1/3	2/3			
kainite	1/5	2/5	2/5		

From the equations

$$F_{K_2}{}^{M} = F_{K_2}{}^{ca} \cdot X^{ca} + F_{K_2}{}^{ka} \cdot X^{ka} + F_{K_2}{}^{R} \cdot X^{R}$$

$$F_{Mg}{}^{M} = F_{Mg}{}^{ca} \cdot X^{ca} + F_{Mg}{}^{ka} \cdot X^{ka} + F_{Mg}{}^{R} \cdot X^{R}$$

$$F_{SO_4}{}^{M} = F_{SO_4}{}^{ka} \cdot X^{ka} + F_{SO_4}{}^{R} \cdot X^{R}$$

or

$$0.2280 = \frac{1}{3} X^{ca} + \frac{1}{5} X^{ka} + 0.0148 \cdot X^{R}$$

$$0.5535 = \frac{2}{3} X^{ca} + \frac{2}{5} X^{ka} + 0.9195 \cdot X^{R}$$

$$0.2185 = \frac{2}{5} X^{ka} + 0.0657 \cdot X^{R}$$

it is calculated that $X^R = 0.1096$, $X^{ka} = 0.5282$, and $X^{ca} = 0.3622$.

The concentration with respect to the projection $C_P = 63.6$ yields

moles	carnallite	kainite	solution R
K_2^{2+}	7.68	6.72	0.10
Mg^{2+}	15.36	13.44	6.41
SO_4^{2-}		13.44	0.46

The quantity of the precipitated halite is obtained from the balance

moles: $Na_2^{2+(M)} = Na_2^{2+(ha)} + Na_2^{2+(R)}$

where

$$Na_2^{2+(M)} = F_{Na_2}{}^{M} \cdot C_P = 24.3$$

and

$$Na_2^{2+(R)} = F_{Na_2}{}^{R} \cdot X^{R} \cdot C_P = 0.18$$

so that

$$Na_2^{2+(ha)} = 24.12.$$

The quantities obtained upon evaporating solution M to dryness are calculated from the equations

$$F_{K_2}{}^{M} = F_{K_2}{}^{ca} \cdot X^{ca} + F_{K_2}{}^{ka} \cdot X^{ka}$$

$$F_{Mg}{}^{M} = F_{Mg}{}^{ca} \cdot X^{ca} + F_{Mg}{}^{ka} \cdot X^{ka} + F_{Mg}{}^{ks} \cdot X^{ks}$$

$$F_{SO_4}{}^{M} = F_{SO_4}{}^{ka} \cdot X^{ka} + F_{SO_4}{}^{ks} \cdot X^{ks}$$

or

$$0.2280 = \frac{1}{3}X^{ca} + \frac{1}{5}X^{ka}$$

$$0.5535 = \frac{2}{3}X^{ca} + \frac{2}{5}X^{ka} + \frac{1}{2}X^{ks}$$

$$0.2185 = \frac{2}{5}X^{ka} + \frac{1}{2}X^{ks}$$

which yield $X^{ca} = 0.5025, X^{ka} = 0.3025$, and $X^{ks} = 0.1950$.

From the concentration with respect to the projection $C_P = 63.6$ it is calculated that

moles	carnallite	kainite	kieserite
K_2^{2+}	10.65	3.85	
Mg^{2+}	21.30	7.70	6.2
SO_4^{2-}		7.70	6.2

The quantity of the precipitated halite is

moles: $Na_2^{2+(ha)} = F_{Na_2}{}^{M} \cdot C_P = 24.3$.

II.6.9.5 Incongruent Dissolution of Bloedite and Carnallite

Adding at 35 °C a solution saturated with respect to halite to an excess of bloedite, glaserite, and halite leads to the formation of thenardite and solution S. The compositions of the solutions are

	$\{Na_2Cl_2\}$	$\{K_2Cl_2\}$	$\{Na_2SO_4\}$	$\{MgSO_4\}$
solution 1	56.0	0	0	0
solution S	41.6	11.4	3.0	18.1

The quantities of the solids and solution S are calculated from the equations

moles	sol. S		ha		th		bl		gs		sol. 1
Na_2^{2+}	44.6	=	u	+	v	+	w	+	$0.5 \cdot x$	+	$56.0 \cdot y$
K_2^{2+}	11.4	=							$1.5 \cdot x$		
Mg^{2+}	18.1	=					w				
Cl_2^{2-}	53.0	=	u							+	$56.0 \cdot y$
SO_4^{2-}	21.1	=			v	+	$2 \cdot w$	+	$2 \cdot x$		
H_2O	10^3	=					$4 \cdot w$			+	$10^3 \cdot y$

$u = 1.05, v = -30.3, w = 18.1, x = 7.6, y = 0.9276.$

It appears that 2.1 mol halite, 18.1 mol bloedite, and 7.6 mol glaserite are dissolved whereas 30.3 mol thenardite are precipitated.

The dissolution of an excess of carnallite and kainite in a solution saturated with respect to halite at 35 °C may yield solution Q. The compositions of the solutions and solids are given in terms of a projection from the Na_2Cl_2 point

	F_{K_2}	F_{Mg}	F_{SO_4}	C_P	F_{Na_2}
solution 1					0.6512
solution Q	0.0767	0.8721	0.0512	86.0	0.05233
carnallite	1/3	2/3			
kainite	1/5	2/5	2/5		

From the equations

$$F_{K_2}{}^{Q} = F_{K_2}{}^{ca} \cdot X^{ca} + F_{K_2}{}^{ka} \cdot X^{ka} + F_{K_2}{}^{sy} \cdot X^{sy}$$

$$F_{Mg}{}^{Q} = F_{Mg}{}^{ca} \cdot X^{ca} + F_{Mg}{}^{ka} \cdot X^{ka}$$

$$F_{SO_4}{}^{Q} = F_{SO_4}{}^{ka} \cdot X^{ka}$$

or

$$0.0767 = \frac{1}{3}X^{ca} + \frac{1}{5}X^{ka} + X^{sy}$$

$$0.8721 = \frac{2}{3}X^{ca} + \frac{2}{5}X^{ka}$$

$$0.0512 = \frac{2}{5}X^{ka}$$

it is calculated that $X^{ka} = 0.1280$, $X^{ca} = 1.2314$, and $X^{sy} = -0.3594$.
It appears that kainite and carnallite are dissolved whereas sylvite is precipitated.

Multiplying the above equations by $C_p = 86.0$ yields

	solution Q	carnallite	kainite	sylvite
K_2^{2+}	6.6	35.3	2.2	–30.9
Mg^{2+}	75.0	70.6	4.4	
SO_4^{2-}	4.4		4.4	

Introducing sodium and chloride of solution Q ($Na_2^{2+} = F_{Na_2}{}^{Q} \cdot C_P = 4.5$, $Cl_2^{2-} = Na_2^{2+} + K_2^{2+} + Mg^{2+} - SO_4^{2-} = 81.7$) and of solution 1 ($Na_2^{2+} = Cl_2^{2-} = F_{Na_2}{}^{sol.1} \cdot C_P = 56.0$) yields the balances

	sol.Q		ca		ka		sy		ha		sol.1
K_2^{2+}	6.6	=	35.3	+	2.2	−	30.9				
Mg^{2+}	75.0	=	70.6	+	4.4						
SO_4^{2-}	4.4	=			4.4						
Na_2^{2+}	4.5	=							u	+	$56.0 \cdot v$
Cl_2^{2-}	81.7	=	105.9	+	2.2	−	30.9	+	u	+	$56.0 \cdot v$
H_2O	10^3	=	423.6	+	12.1					+	$10^3 \cdot v$

$u = -27.1, v = 0.5643.$

In addition to sylvite a quantity of 54.2 mol halite has been precipitated.

II.6.9.6 Generation of a Solution from Hydrated Solids

At 8.5 °C the stable solids are carnallite, epsomite, kainite, sylvite, and halite. As kainite is less hydrated than carnallite and epsomite a solution saturated with respect to the above solids will be formed if a dry mixture of carnallite + epsomite + sylvite + halite is heated to a temperature of 8.5 °C. The composition of the solution is

$\{Na_2Cl_2\}$	$\{K_2Cl_2\}$	$\{MgCl_2\}$	$\{MgSO_4\}$
4.4	4.2	64.6	6.4

The quantities of the solids are obtained from the equations

	solution		ha		sy	ep		ca		ka
Na_2^{2+}	4.4	=	u							
K_2^{2+}	4.2	=			v		+	$0.5 \cdot x$	+	$0.5 \cdot y$
Mg^{2+}	71.0	=				w	+	x	+	y
Cl_2^{2-}	73.2	=	u	+	v		+	$1.5 \cdot x$	+	$0.5 \cdot y$
SO_4^{2-}	6.4	=				w			+	y
H_2O	10^3	=				$7 \cdot w$	+	$6.0 \cdot x$	+	$2.75 \cdot y$

$u = 4.4, v = 38.68, w = 139.95, x = 64.6, y = -133.55.$

It appears that 8.8 mol halite + 77.36 mol sylvite + 139.95 mol epsomite + 64.6 mol carnallite will react to 133.55 mol kainite and a saturated solution.

At 69.0 °C carnallite + kainite + halite will react to kieserite + sylvite and a saturated solution. The quantities of the solids are obtained from the equations

	solution		ha		sy	ks		ca		ka
Na_2^{2+}	4.3	=	u							
K_2^{2+}	10.1	=			v		+	$0.5 \cdot x$	+	$0.5 \cdot y$
Mg^{2+}	83.3	=				w	+	x	+	y
Cl_2^{2-}	94.7	=	u	+	v		+	$1.5 \cdot x$	+	$0.5 \cdot y$
SO_4^{2-}	3.0	=				w			+	y
H_2O	10^3	=				w	+	$6.0 \cdot x$	+	$2.75 \cdot y$

$u = 4.3, v = -177.25, w = -291.4, x = 80.3, y = 294.4.$

At 78.0 °C kainite + halite will react to kieserite + langbeinite + sylvite and a saturated solution. The composition of the solution is given in terms of a projection from the Na_2Cl_2 point

F_{K_2}	F_{Mg}	F_{SO_4}	C_P	F_{Na_2}
0.1610	0.7850	0.0540	90.7	0.0827

From the equations

$$F_{K_2}{}^{sol.} = F_{K_2}{}^{sy} \cdot X^{sy} + F_{K_2}{}^{ka} \cdot X^{ka} + F_{K_2}{}^{lg} \cdot X^{lg}$$

$$F_{Mg}{}^{sol} = F_{Mg}{}^{ka} \cdot X^{ka} + F_{Mg}{}^{lg} \cdot X^{lg} + F_{Mg}{}^{ks} \cdot X^{ks}$$

$$F_{SO_4}{}^{sol} = F_{SO_4}{}^{ka} \cdot X^{ka} + F_{SO_4}{}^{lg} \cdot X^{lg} + F_{SO_4}{}^{ks} \cdot X^{ks}$$

it follows that there are four unknown quantities so that the F–values have to be transformed into concentrations. The concentration of the solution is

$\{Na_2Cl_2\}$	$\{K_2Cl_2\}$	$\{MgCl_2\}$	$\{MgSO_4\}$
7.5	14.6	66.3	4.9

The quantities of the solids are calculated from the equations

	solution		ha		sy	ks		ka		lg
Na_2^{2+}	7.5	=	u							
K_2^{2+}	14.6	=			v		+	$0.5 \cdot x$	+	y
Mg^{2+}	71.2	=				w	+	x	+	$2 \cdot y$
Cl_2^{2-}	88.4	=	u	+	v		+	$0.5 \cdot x$		
SO_4^{2-}	4.9	=				w	+	x	+	$3 \cdot y$
H_2O	10^3	=				w	+	$2.75 \cdot x$		

$u = 7.5, v = -146.59, w = -251.17, x = 454.97, y = -66.3.$

It appears that sylvite, kieserite, and langbeinite are produced whereas halite and kainite are consumed.

Suggested Reading

Althammer, W.: Die graphische und rechnerische Behandlung von Salzlösungen. Kali-Forschungs-Anstalt GmbH, Staßfurt-Leopoldshall (1924), 55 pp

Blasdale, W. C.: Equilibria in saturated salt solutions. Am. Chem. Soc. Monogr. Ser. The Chemical Catalog Co., New York (1927), 197 pp

Findlay, A.: The phase rule and its application. Dover Publications, New York (1951), 368 pp. German edition: Die Phasenregel und ihre Anwendungen. Verlag Chemie, Weinheim/Bergstr. (1958), 367 pp.

Jänecke, E.: Neue Darstellungsform der wässerigen Lösungen. Z. anorg. Chem. 51:132-157 (1906)

Jänecke, E.: Neue Darstellungsform der van't Hoffschen Untersuchungen II. Z. anorg. Chem. 52:358-367 (1907)

Jänecke, E.: Über eine Darstellungsform der van't Hoffschen Untersuchungen über ozeanische Salzablagerungen III. Z. anorg. Chem. 53:319-326 (1907)

Jänecke, E.: Gesättigte Salzlösungen vom Standpunkt der Phasenlehre. Knapp, W., Halle a. d. Saale (1908), 188 pp

Jänecke, E.: Über reziproke Salzpaare und doppelt-ternäre Salzmischungen. Z. phys. Chem. 82:1-34 (1913)

III Data and Diagrams for Solid-Solution Equilibria

III.1
Compositions and Concentrations

III.1.1
Minerals and Seawater

Table 3.1 compiles in the upper part the compositions of minerals which are treated in the following sections. Their compositions in terms of projections are given in the sections concerned with the pertinent systems. Other important minerals which occur in salt deposits are listed in the lower part of the table. Table 3.2 shows the composition of modern seawater.

III.1.2
Concentration Units

By convention the compositions of highly concentrated salt solutions are expressed by the molar salt to water ratio, R, or by the weight ratio of salt to water, Q. The unit of R is "moles anhydrous salt (solute) dissolved in 1000 mol of water". This unit is denoted by the curley brackets { }. The unit of Q is "grams of anhydrous salt dissolved in 100 grams of water", which is denoted as g/a.
Using the notations

S : moles of solute
F_S : formula weight of solute
$G_S = S \cdot F_S$: weight of solute
W : moles of water
F_W : formula weight of water
$G_W = W \cdot F_W$: weight of water

the molar and weight ratios of salt to water are expressed by the equations:

$$R = \frac{S}{W} \cdot 10^3 \qquad \text{(moles of solute per } 10^3 \text{ mol of water)} \qquad \text{(III.1)}$$

$$Q = \frac{G_S}{G_W} \cdot 10^2 \qquad \text{(grams of solute per } 10^2 \text{ grams of water)} \qquad \text{(III.2)}$$

Both quantities are interrelated by the expression

$$\frac{R}{Q} = \frac{F_W}{F_S} \cdot 10 \qquad \text{(III.3)}$$

Conversion factors are

	R → Q	Q → R		R → Q	Q → R
Na_2Cl_2	0.6488	1.5412	Na_2SO_4	0.7885	1.2683
K_2Cl_2	0.8277	1.2082	K_2SO_4	0.9674	1.0337
$MgCl_2$	0.5285	1.8920	$MgSO_4$	0.6682	1.4966

The concentration units molality, m, formality, f, and molarity, M, are defined as

$$m = \frac{S}{G_W} \cdot 10^3 \qquad \text{(moles of solute per } 10^3 \text{ grams of water)} \qquad \text{(III.4)}$$

$$f = \frac{S}{G_L} \cdot 10^3 \qquad \text{(moles of solute per } 10^3 \text{ grams of solution)} \qquad \text{(III.5)}$$

where $G_L = G_S + G_W$ is the weight of the solution, and

$$M = \frac{S}{V_L} \cdot 10^3 \qquad \text{(moles of solute per liter of solution)} \qquad \text{(III.6)}$$

where V_L is the volume of the solution, which is related to G_L by the density of the solution

$$d_L = \frac{G_L}{V_L} \qquad \text{(III.7)}$$

The quantities R and Q are related to m, f, and M by the expressions

$$\frac{R}{m} = F_W \qquad \text{(III.8)}$$

$$\frac{Q}{m} = F_S \cdot 10^{-1} \qquad \text{(III.9)}$$

$$\frac{R}{f} = \frac{G_L \cdot F_W}{G_W} \qquad \text{(III.10)}$$

$$\frac{Q}{f} = \frac{G_L \cdot F_S}{G_W} \cdot 10^{-1} \qquad \text{(III.11)}$$

$$\frac{R}{M} = \frac{G_L \cdot F_W}{G_W \cdot d_L} \tag{III.12}$$

$$\frac{Q}{M} = \frac{G_L \cdot F_S}{G_W \cdot d_L} \cdot 10^{-1} \tag{III.13}$$

It may be seen that R and Q may be readily transformed into molalities and formalities, and vice versa, if either one of them is known. However, these quantities cannot be transformed into molarities, and vice versa, unless the density of the solution is known. With respect to rather dilute solutions the density may be taken as that of the solvent. But it differs distinctly from that of the solvent with respect to highly concentrated solutions. Unfortunately, the densities of the solutions under consideration are not as well known as is desirable. Some examples for saturated solutions are

g/cm^3	Na_2Cl_2	K_2Cl_2	$MgCl_2$	Na_2SO_4	K_2SO_4	$MgSO_4$
25°C	1.200	1.180	1.343	1.211	1.086	1.306
75°C	1.178	1.203	1.436	1.278	1.116	1.432

Table 3.1 Composition of minerals

		Formula weight
ac: arcanite	K_2SO_4	174.27
bi: bischofite	$MgCl_2{\cdot}6H_2O$	203.31
bl: bloedite	$Na_2Mg(SO_4)_2{\cdot}4H_2O$	334.47
ca: carnallite	$KMgCl_3{\cdot}6H_2O$	277.86
da: d'ansite	$Na_{21}MgCl_3(SO_4)_{10}$	1574.08
ep: epsomite	$MgSO_4{\cdot}7H_2O$	246.48
gs: glaserite	$K_3Na(SO_4)_2$	332.42
ha: halite	$NaCl$	58.44
hx: hexahydrite	$MgSO_4{\cdot}6H_2O$	228.46
ka: kainite	$KMgClSO_4{\cdot}11/4\ H_2O$	244.47
ks: kieserite	$MgSO_4{\cdot}H_2O$	138.39
lg: langbeinite	$K_2Mg_2(SO_4)_3$	415.01
le: leonite	$K_2Mg(SO_4)_2{\cdot}4H_2O$	366.70
lw: loeweite	$Na_{12}Mg_7(SO_4)_{13}{\cdot}15H_2O$	1965.09
mi: mirabilite	$Na_2SO_4{\cdot}10H_2O$	322.19
pc: picromerite	$K_2Mg(SO_4)_2{\cdot}6H_2O$	402.73
sy: sylvite	KCl	74.56
th: thenardite	Na_2SO_4	142.04
vh: vanthoffite	$Na_6Mg(SO_4)_4$	546.49
anhydrite	$CaSO_4$	136.14
chlorocalcite	$KCaCl_3$	185.54
gypsum	$CaSO_4{\cdot}2H_2O$	172.17
glauberite	$Na_2Ca(SO_4)_2$	278.18
goergeyite	$K_2Ca_5(SO_4)_6{\cdot}H_2O$	872.99
polyhalite	$Ca_2K_2Mg(SO_4)_4{\cdot}2H_2O$	602.95
syngenite	$K_2Ca(SO_4)_2{\cdot}H_2O$	328.42
tachhydrite	$CaMg_2Cl_6{\cdot}12H_2O$	517.60

Table 3.2 Composition of seawater (Culkin 1967)

		Moles		Fictious components		
	g/kg	10^3 mol	H_2O		mol%	wt%
Cl^-	19.353	Cl_2^{2-}	5.096	NaCl	86.5	78.3
Na^+	10.760	Na_2^{2+}	4.369	$MgCl_2$	6.1	8.9
SO_4^{2-}	2.712	SO_4^{2-}	0.527	$MgSO_4$	3.7	6.9
Mg^{2+}	1.294	Mg^{2+}	0.994	KCl	1.8	2.1
Ca^{2+}	0.413	Ca^{2+}	0.192	$CaSO_4$	1.5	3.1
K^+	0.387	K_2^{2+}	0.092	$CaCO_3$	0.4	0.7
HCO_3^-	0.142	HCO_3^-	0.043			

III.2.1
Na_2Cl_2-H_2O

ha: halite (NaCl), hh: hydrohalite ($NaCl \cdot 2H_2O$)

	°C	$\{Na_2Cl_2\}$
b: ice + hh	−21.2	47.3
c: hh + ha	0.2	54.8

SALSYS
Input: temperature. Output: $\{Na_2Cl_2\}$

		°C	
a–b: ice	0	–	−21.2
b–c: hh	−21.2	–	0.2
c–d: ha	0.2	–	210.0

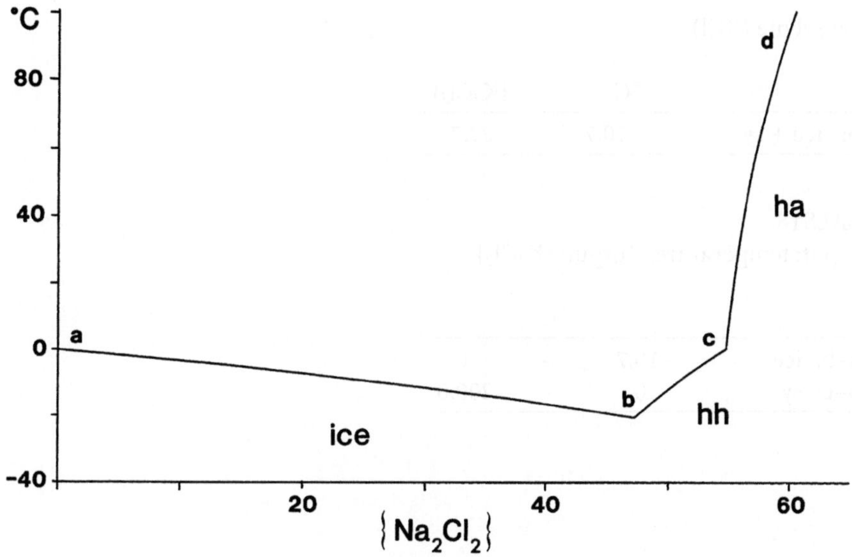
°C
80
40
0
-40
20
40
60
{Na2Cl2}
a
b
c
d
ice
hh
ha

III.2.2
K_2Cl_2-H_2O

sy: sylvite (KCl)

	°C	$\{K_2Cl_2\}$
b: ice + sy	−10.7	29.7

SALSYS
Input: temperature. Output: $\{K_2Cl_2\}$

	°C		
a–b: ice	−10.7	–	0
b–c: sy	0	–	200.0

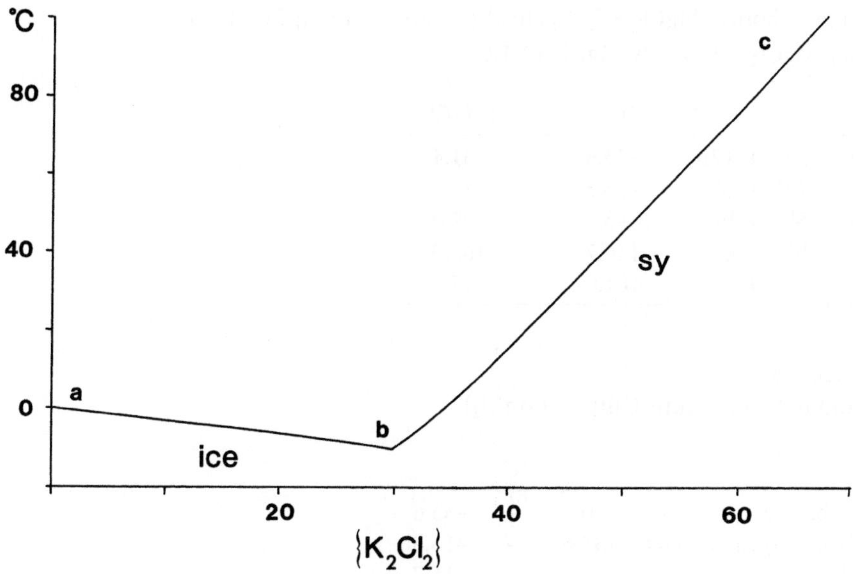
°C
80
40
0
a
b
c
ice
sy
20
40
60
{K2Cl2}

III.2.3
$MgCl_2$-H_2O

bi: bischofite ($MgCl_2 \cdot 6H_2O$), 2h: $MgCl_2 \cdot 2H_2O$, 4h: $MgCl_2 \cdot 4H_2O$
8h: $MgCl_2 \cdot 8H_2O$, 12h: $MgCl_2 \cdot 12H_2O$

		°C	$\{MgCl_2\}$
b:	ice + 12h	−33.6	51.4
c:	12h + 8h	−16.8	87.5
d:	8h + bi	−3.4	99.0
e:	bi + 4h	116.7	162.0
f:	4h + 2h	181.0	217.4

SALSYS
Input: temperature. Output: $\{MgCl_2\}$

			°C	
a–b:	ice	0	-	−33.6
b–c:	$MgCl_2 \cdot 12H_2O$	−33.6	-	−16.8
d–e:	bi	−3.4	-	116.7
e–f:	$MgCl_2 \cdot 4H_2O$	116.7	-	181.0

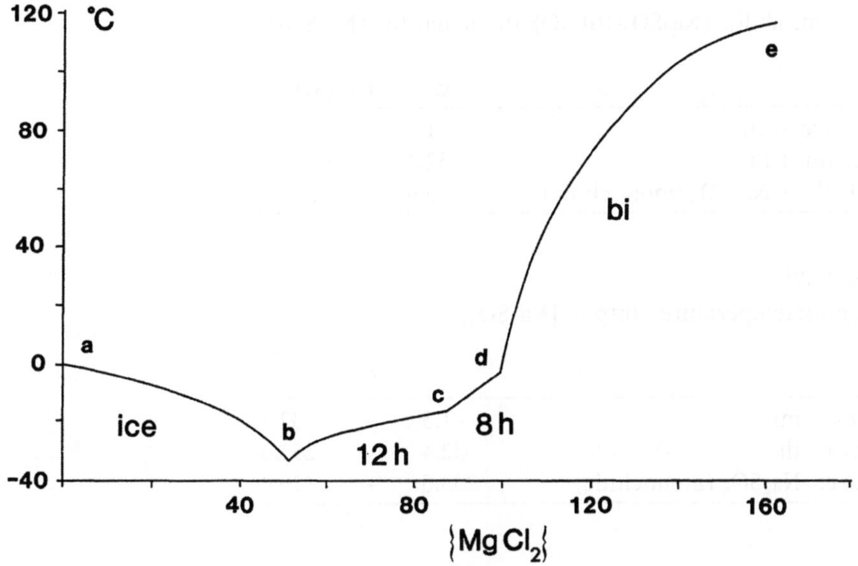
120
°C
80
40
0
-40
a
b
c
d
e
ice
12 h
8 h
bi
40
80
120
160
{MgCl$_2$}

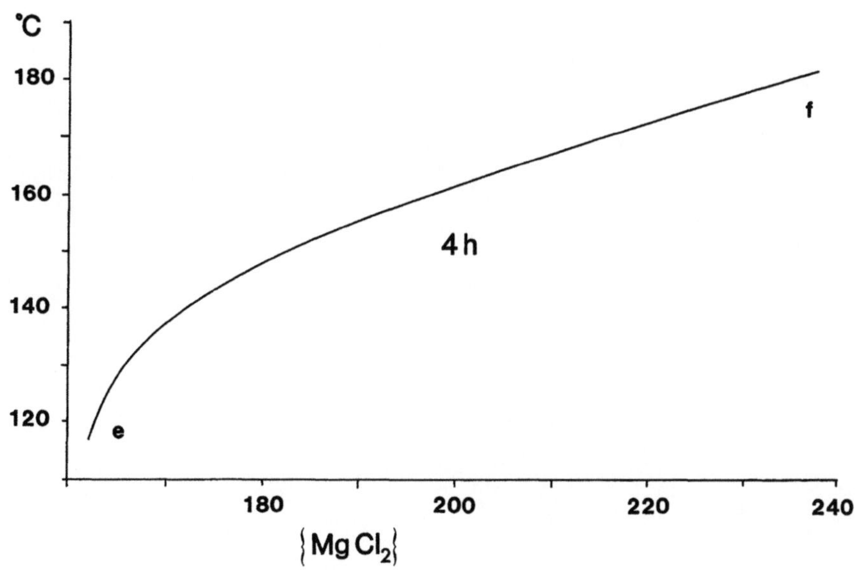
°C
180
160
140
120
e
f
4 h
180
200
220
240
{MgCl$_2$}

III.2.4
Na_2SO_4-H_2O

mi: mirabilite ($Na_2SO_4 \cdot 10H_2O$), th: thenardite (Na_2SO_4)

	°C	$\{Na_2SO_4\}$
b: ice + mi	−1.3	5.4
c: mi + th	32.4	63.1
d: th + Na_2SO_4 (monoclinic)	233.0	59.5

SALSYS
Input: temperature. Output: $\{Na_2SO_4\}$

	°C		
b–c: mi	−1.3	–	32.4
c–d: th	32.4	–	233.0
d–e: Na_2SO_4 (monoclinic)	233.0	–	320.0

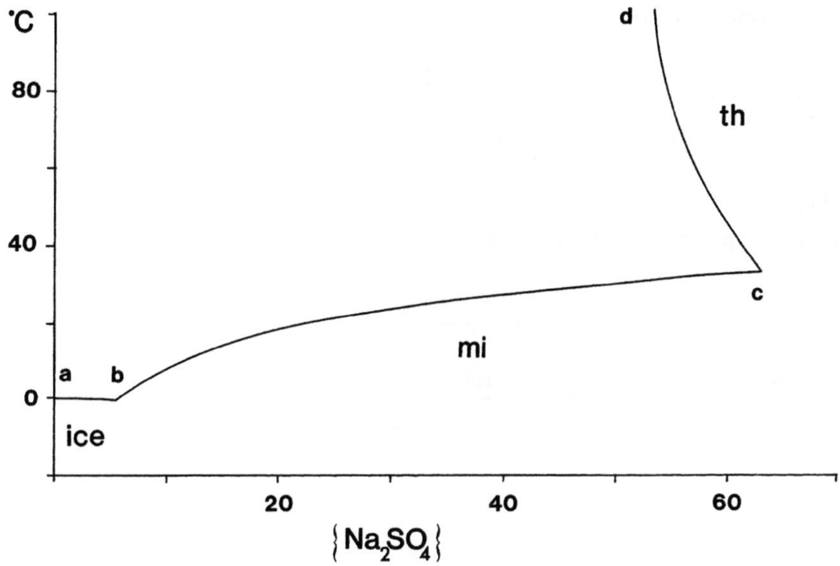
°C
80
40
0
d
th
c
mi
a
b
ice
20
40
60
{Na2SO4}

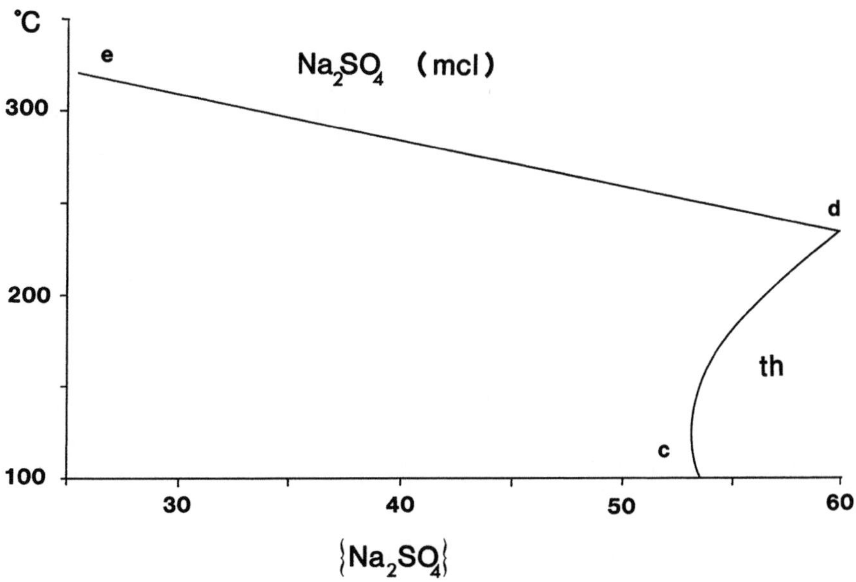
°C
300
200
100
e
Na2SO4 (mcl)
d
th
c
30
40
50
60
{Na2SO4}

III.2.5
K_2SO_4-H_2O

ac: arcanite (K_2SO_4)

	°C	$\{K_2SO_4\}$
b: ice + ac	−1.5	7.3

SALSYS
Input: temperature. Output: $\{K_2SO_4\}$

	°C		
a–b: ice	−1.5	–	0
b–c: ac	−1.5	–	190

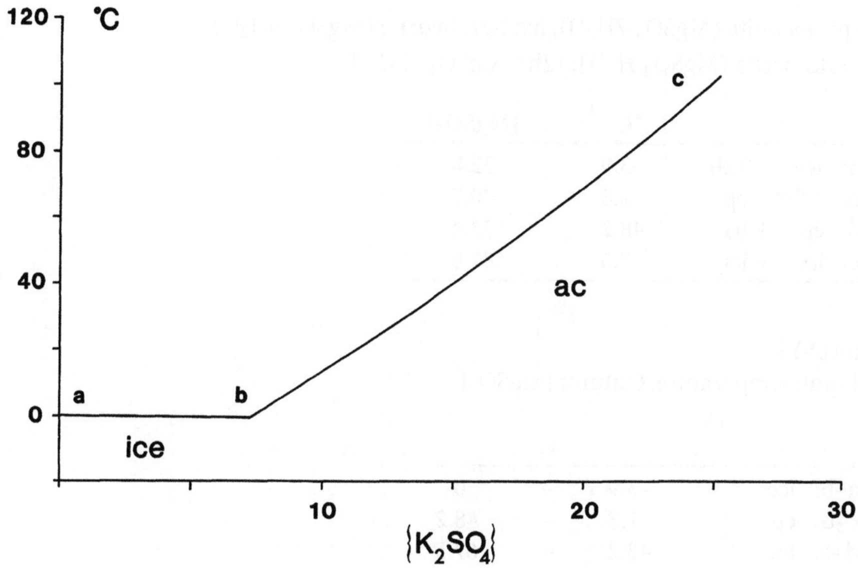

120
°C
80
40
0
a
b
c
ac
ice
10
20
30
{K2SO4}

III.2.6
$MgSO_4$-H_2O

ep: epsomite ($MgSO_4 \cdot 7H_2O$), hx: hexahydrite ($MgSO_4 \cdot 6H_2O$)
ks: kieserite ($MgSO_4 \cdot H_2O$), 12h*: $MgSO_4 \cdot 12H_2O$

	°C	{$MgSO_4$}
b: ice + 12h*	–3.9	32.8
c: 12h* + ep	1.8	40.7
d: ep + hx	48.2	73.4
e: hx + ks	67.5	87.8

SALSYS
Input: temperature. Output: {$MgSO_4$}

		°C		
a–b:	ice	–3.9	–	0
c–d:	ep	1.8	–	48.2
d–e:	hx	48.2	–	67.5
e–f:	ks	67.5	–	190.0

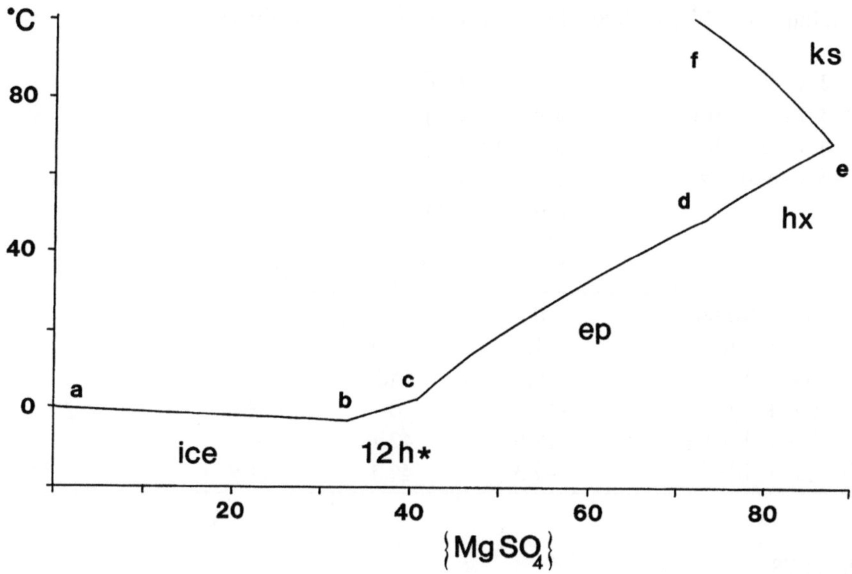
°C
80
40
0
a
b
c
d
e
f
ks
hx
ep
ice
12 h*
20
40
60
80
{Mg SO4}

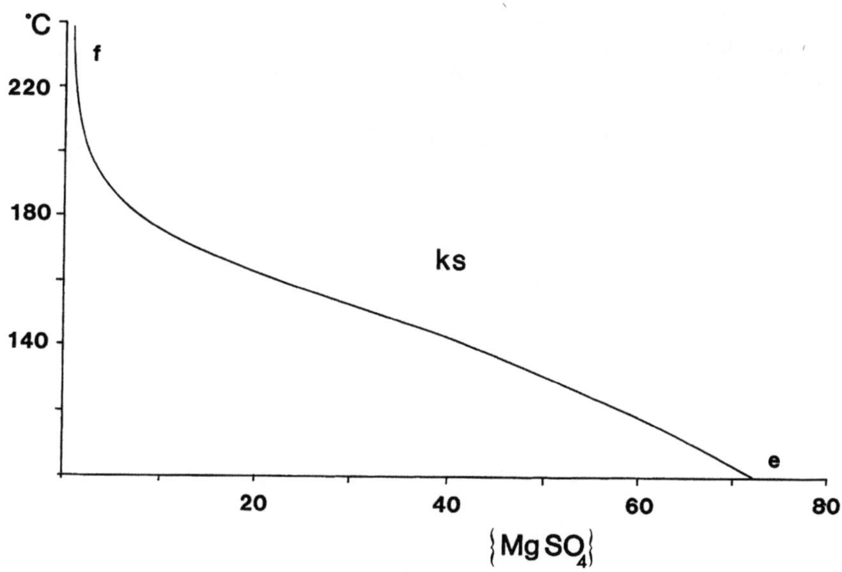
°C
220
180
140
f
ks
e
20
40
60
80
{Mg SO4}

III.3.1
Na_2Cl_2-K_2Cl_2-H_2O

ha: halite (NaCl), hh: hydrohalite ($NaCl{\cdot}2H_2O$), sy: sylvite (KCl)

1–2–a–5:	ice	a–2: ice + hh
2–3–b–a:	hh	a–5: ice + sy
3–4–c–b:	ha	a–b: hh + sy
5–6–c–b–a:	sy	b–3: ha + hh
		b–c: ha + sy

	°C	$\{Na_2Cl_2\}$	$\{K_2Cl_2\}$
1: ice + water	0	0	0
2: ice + hh	−21.2	47.3	0
3: ha + hh	0.2	54.8	0
5: ice + sy	−10.7	0	29.7
a: ice + hh + sy	−22.9	42.1	9.4
b: ha + hh + sy	−2.3	49.3	11.9

SALSYS
Input: temperature. Output: $\{Na_2Cl_2\}$, $\{K_2Cl_2\}$

		°C	
a–2: ice + hh	−22.9	–	−21.2
a–5: ice + sy	−22.9	–	−10.7
a–b: hh + sy	−22.9	–	−2.3
b–c: ha + sy	−2.3	–	190.0

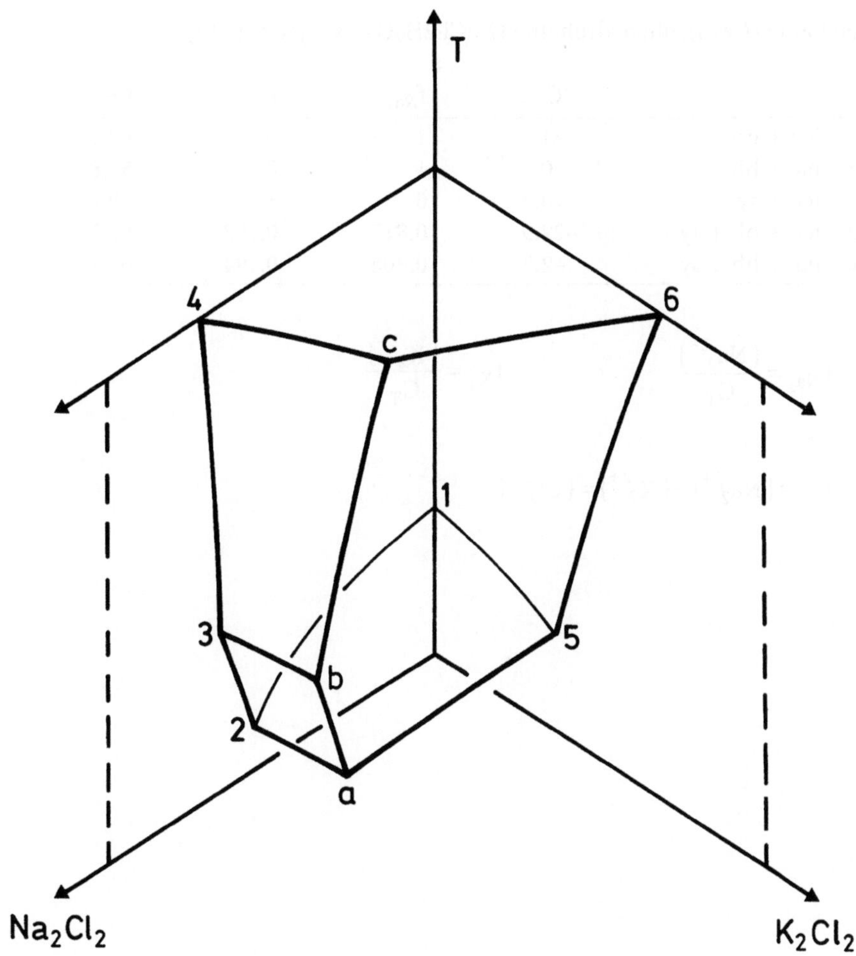
T
4
6
c
1
3
5
b
2
a
Na_2Cl_2
K_2Cl_2

III.3.1
Na_2Cl_2-K_2Cl_2-H_2O

ha: halite (NaCl), hh: hydrohalite ($NaCl{\cdot}2H_2O$), sy: sylvite (KCl)

	°C	f_{Na_2}	f_{K_2}	C_T
2: ice + hh	−21.2	1	0	47.3
3: ha + hh	0.2	1	0	54.8
5: ice + sy	−10.7	0	1	29.7
a: ice + hh + sy	−22.9	0.817	0.183	51.5
b: ha + hh + sy	−2.3	0.806	0.194	61.2

$$f_{Na_2} = \frac{\{Na_2^{2+}\}}{C_T} \qquad f_{K_2} = \frac{\{K_2^{2+}\}}{C_T}$$

$$C_T = \{Na_2^{2+}\} + \{K_2^{2+}\} = \{Cl_2^{2-}\}$$

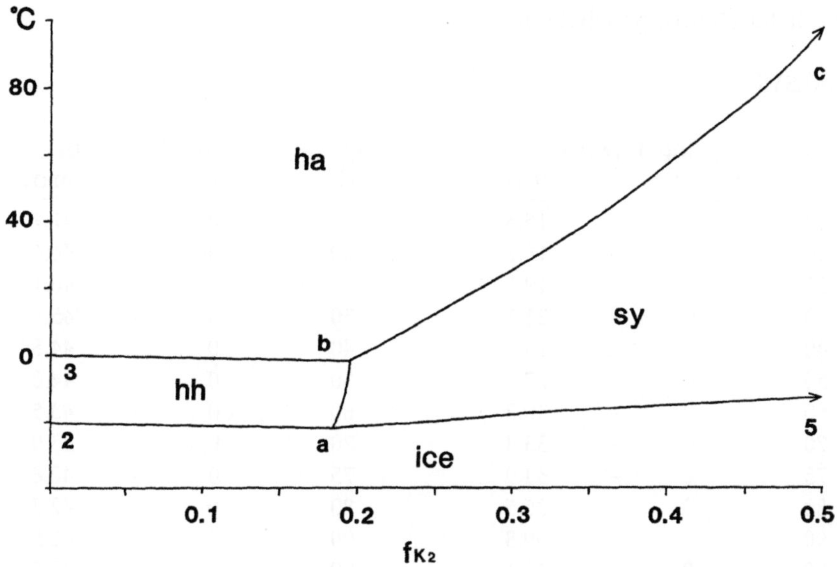
°C
80
40
0
0.1
0.2
0.3
0.4
0.5
fK2
ha
hh
sy
ice
a
b
c
2
3
5

III.3.1
Na_2Cl_2-K_2Cl_2-H_2O

ha: halite (NaCl), sy: sylvite (KCl)

SALSYS

ha	Input: $\{K_2Cl_2\}$			sy	Input: $\{Na_2Cl_2\}$		
°C	min.		max.	°C	min.		max.
10	0	–	15.3	10	0	–	47.7
20	0	–	18.2	20	0	–	46.6
25	0	–	19.7	25	0	–	46.1
30	0	–	21.1	30	0	–	45.6
40	0	–	24.1	40	0	–	44.8
50	0	–	27.1	50	0	–	44.0
60	0	–	30.2	60	0	–	43.5
70	0	–	33.4	70	0	–	43.0
75	0	–	34.9	75	0	–	42.8
80	0	–	36.5	80	0	–	42.7
90	0	–	39.8	90	0	–	42.5
100	0	–	43.1	100	0	–	42.5
120	0	–	49.4	120	0	–	42.9
125	0	–	50.9	125	0	–	43.0
140	0	–	55.7	140	0	–	43.5
150	0	–	58.9	150	0	–	44.0
170	0	–	65.4	170	0	–	45.1
190	0	–	72.0	190	0	–	46.5
	Output: $\{Na_2Cl_2\}$				Output: $\{K_2Cl_2\}$		

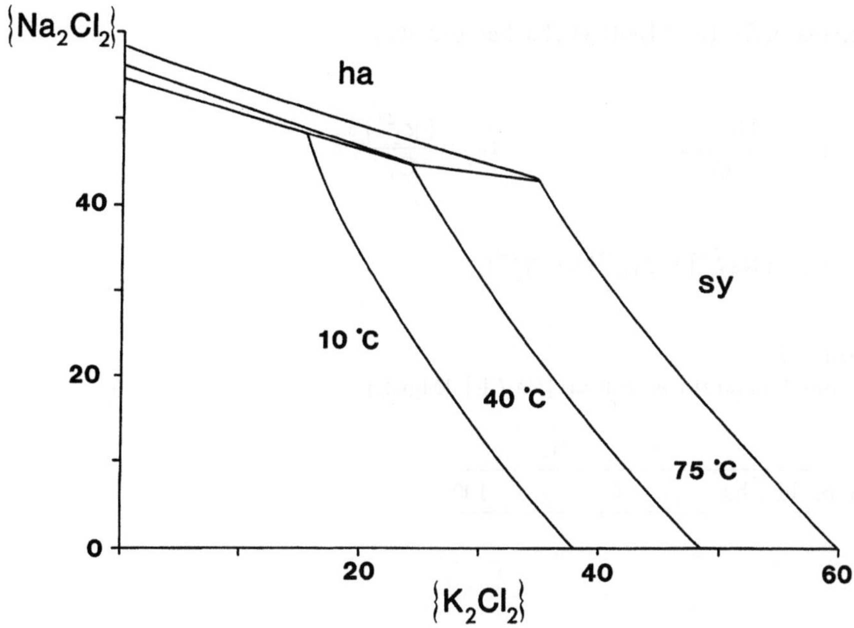
{Na2Cl2}
ha
sy
10 °C
40 °C
75 °C
0
20
40
20
40
60
{K2Cl2}

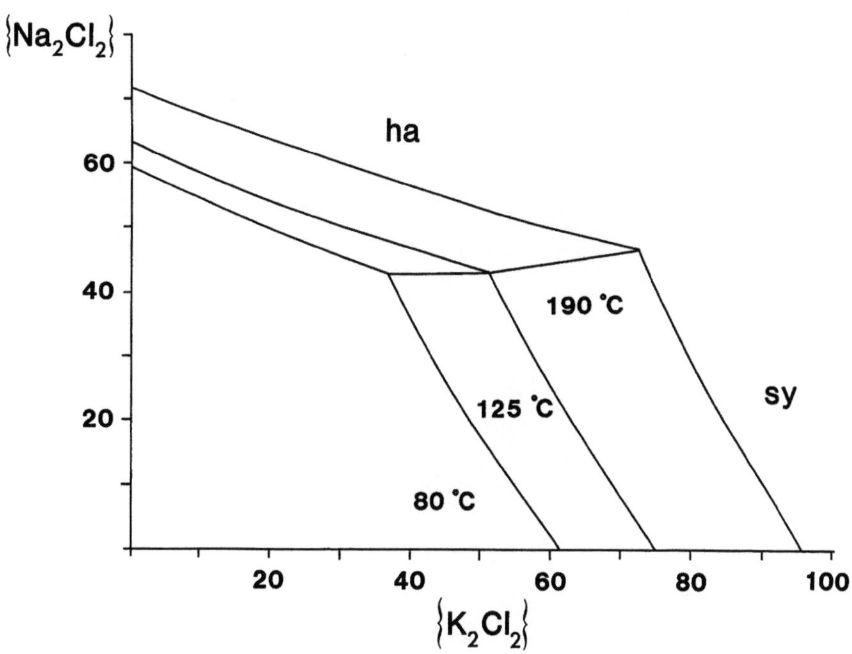
{Na2Cl2}
ha
sy
190 °C
125 °C
80 °C
20
40
60
20
40
60
80
100
{K2Cl2}

III.3.2
Na_2Cl_2-$MgCl_2$-H_2O

bi: bischofite ($MgCl_2 \cdot 6H_2O$), ha: halite (NaCl)

$$f_{Na_2} = \frac{\{Na_2^{2+}\}}{C_T} \qquad f_{K_2} = \frac{\{K_2^{2+}\}}{C_T}$$

$$C_T = \{Na_2^{2+}\} + \{Mg^{2+}\} = \{Cl_2^{2-}\}$$

SALSYS
Input: temperature. Output: $\{Na_2Cl_2\}$, $\{MgCl_2\}$

	°C		
a–b: bi + ha	0	–	100

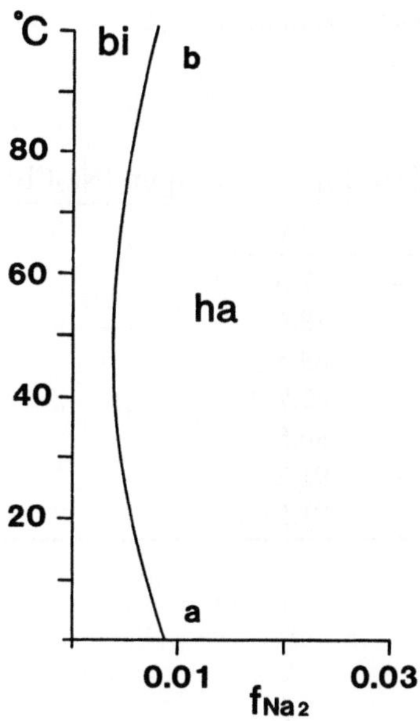

°C
bi
b
80
60
ha
40
20
a
0.01
0.03
f_{Na2}

III.3.2
Na_2Cl_2-$MgCl_2$-H_2O

bi: bischofite ($MgCl_2 \cdot 6H_2O$), ha: halite (NaCl)

SALSYS

ha	Input: $\{MgCl_2\}$			Output: $\{Na_2Cl_2\}$
°C	min.		max.	
12.5	10.0	–	77.3	
25	10.3	–	79.4	
35	10.1	–	89.3	
50	12.8	–	82.7	
55	10.0	–	90.5	
75	10.2	–	91.2	
90	10.3	–	91.5	

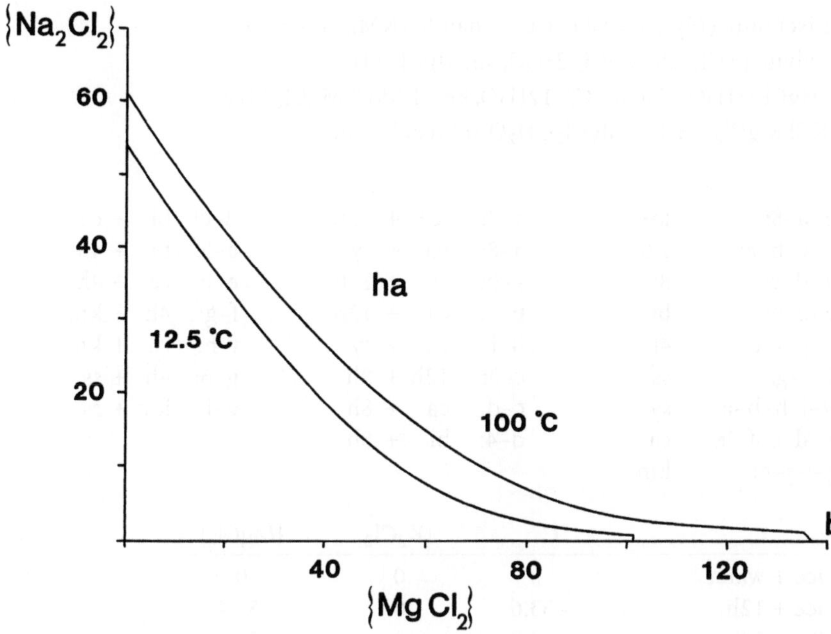
{Na2Cl2}
60
40
20
ha
12.5 °C
100 °C
bi
40
80
120
{MgCl2}

III.3.3
K_2Cl_2-$MgCl_2$-H_2O

bi: bischofite ($MgCl_2 \cdot 6H_2O$), ca: carnallite ($KMgCl_3 \cdot 6H_2O$)
sy: sylvite (KCl), 2h: $MgCl_2 \cdot 2H_2O$, 4h: $MgCl_2 \cdot 4H_2O$
8h: $MgCl_2 \cdot 8H_2O$, 12h: $MgCl_2 \cdot 12H_2O$, km: $1.5KCl \cdot MgCl_2 \cdot 2H_2O$
ss: $KCl \cdot MgCl_2 \cdot 2H_2O$ – $MgCl_2 \cdot 2H_2O$ solid solutions

1–2–a–8:	ice	a–2:	ice + 12h	d–e:	bi + ca
2–3–c–b–a:	12h	a–8:	ice + sy	e–5:	bi + 4h
3–4–d–c:	8h	a–b:	sy + 12h	e–f:	ca + 4h
4–5–e–d:	bi	b–c:	ca + 12h	f–g:	4h + km
5–6–g–f–e:	4h	b–h:	ca + sy	f–h:	ca + km
6–7–i–g:	ss	c–3:	12h + 8h	g–6:	4h + ss
8–9–j–h–b–a:	sy	c–d:	ca + 8h	g–i:	km + ss
b–c–d–e–f–h:	ca	d–4:	bi + 8h		
f–g–i–j–h:	km				

	°C	$\{K_2Cl_2\}$	$\{MgCl_2\}$
1: ice + water	0	0	0
2: ice + 12h	–33.6	0	51.4
3: 8h + 12h	–16.8	0	87.5
4: bi + 8h	–3.4	0	99.0
5: bi + 4h	116.7	0	162.0
6: 2h + 4h	181.0	0	217.4
8: ice + sy	–10.7	29.7	0
a: ice + sy + 12h	–34.3	1.8	51.0
b: ca + sy + 12h	–21.0	2.5	66.1
c: ca + 8h + 12h	–16.9	1.5	87.5
d: bi + ca + 8h	–3.4	0.6	99.0
e: bi + ca + 4h	116.7	2.6	162.0
f: ca + 4h + km	153	35.2	224.8
g: 4h + km + ss	165	36.9	225.9
h: ca + sy + km	157	37.9	212.0

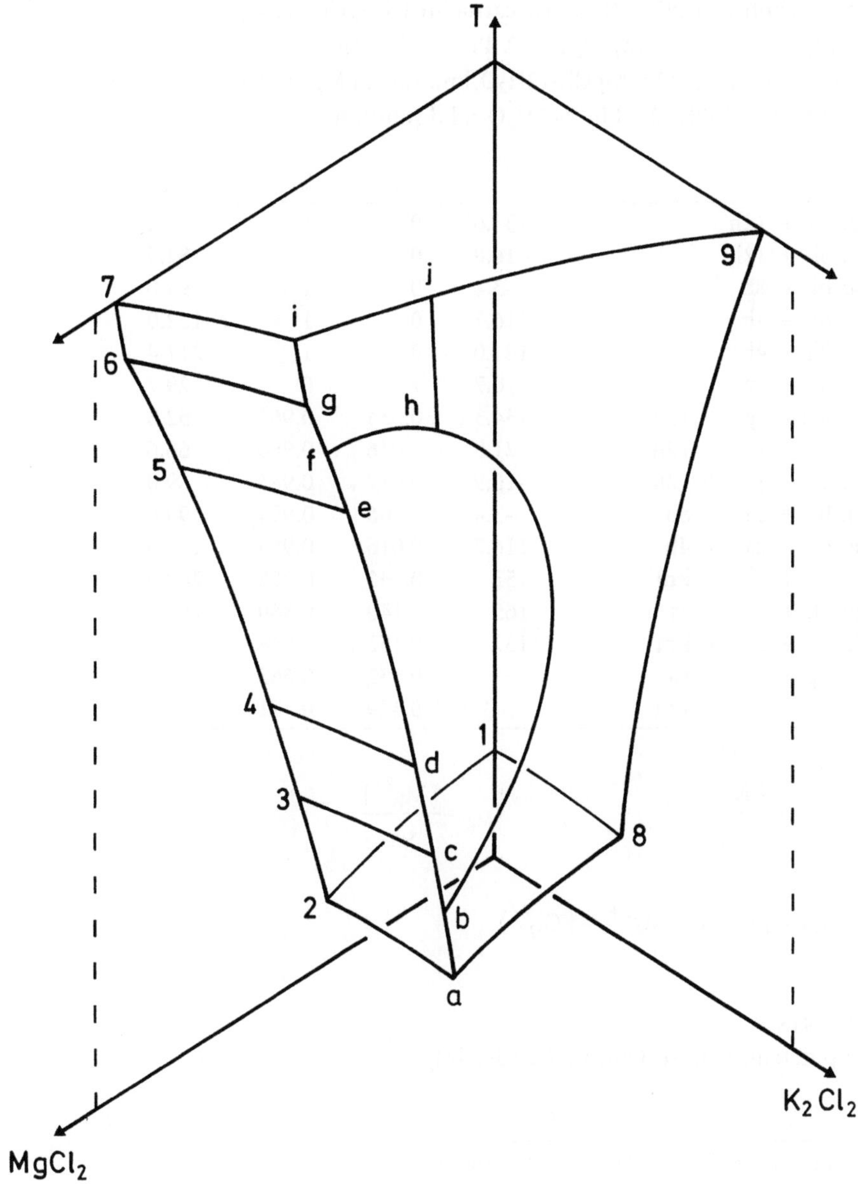
T
7
i
j
9
6
g
h
f
5
e
4
1
d
3
8
c
2
b
a
MgCl$_2$
K$_2$Cl$_2$

III.3.3 K_2Cl_2-$MgCl_2$-H_2O

bi: bischofite ($MgCl_2 \cdot 6H_2O$), ca: carnallite ($KMgCl_3 \cdot 6H_2O$)
sy: sylvite (KCl), 2h: $MgCl_2 \cdot 2H_2O$, 4h: $MgCl_2 \cdot 4H_2O$
8h: $MgCl_2 \cdot 8H_2O$, 12h: $MgCl_2 \cdot 12H_2O$, km: $1.5KCl \cdot MgCl_2 \cdot 2H_2O$
ss: $KCl \cdot MgCl_2 \cdot 2H_2O$ – $MgCl_2 \cdot 2H_2O$ solid solutions

	°C	f_{K_2}	f_{Mg}	C_T
2: ice + 12h	−33.6	0	1	51.4
3: 8h + 12h	−16.8	0	1	87.5
4: bi + 8h	−3.4	0	1	99.0
5: bi + 4h	116.7	0	1	162.0
6: 2h + 4h	181.0	0	1	217.4
8: ice + sy	−10.7	1	0	29.7
a: ice + sy + 12h	−34.3	0.034	0.966	52.8
b: ca + sy + 12h	−21.0	0.036	0.964	68.6
c: ca + 8h + 12h	−16.9	0.017	0.983	89.0
d: bi + ca + 8h	−3.4	0.006	0.994	99.6
e: bi + ca + 4h	116.7	0.016	0.984	164.6
f: ca + 4h + km	153	0.135	0.865	260.0
g: 4h + km + ss	165	0.140	0.860	262.8
h: ca + sy + km	157	0.152	0.848	249.9
salt point: ca		0.333	0.667	
km		0.429	0.571	

$$f_{K_2} = \frac{\{K_2^{2+}\}}{C_T} \qquad f_{Mg} = \frac{\{Mg^{2+}\}}{C_T}$$

$$C_T = \{K_2^{2+}\} + \{Mg^{2+}\} = \{Cl_2^{2-}\}$$

SALSYS
Input: temperature. Output: $\{K_2Cl_2\}$, $\{MgCl_2\}$

		°C		
b–h:	ca + sy	−21.0	–	150.0
d–e:	bi + ca	−3.4	–	105.0
e–f:	ca + 4h	116.7	–	152.0
h–j:	sy + km	157.0	–	220.0

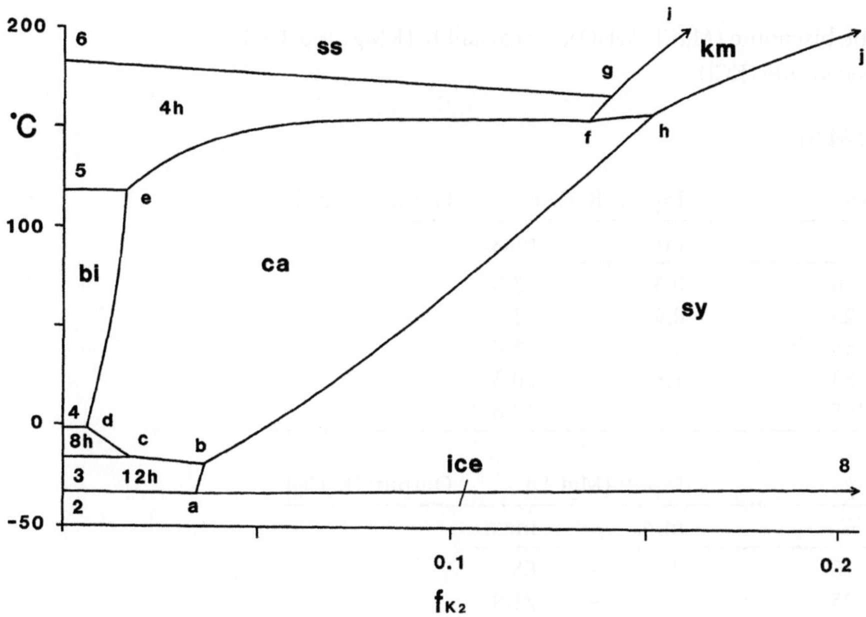

200
100
0
-50
°C
6
5
4
3
2
ss
4h
bi
ca
8h
12h
ice
sy
km
a
b
c
d
e
f
g
h
i
j
8
0.1
0.2
f_{K_2}

III.3.3
K_2Cl_2-$MgCl_2$-H_2O

bi: bischofite ($MgCl_2 \cdot 6H_2O$), ca: carnallite ($KMgCl_3 \cdot 6H_2O$)
sy: sylvite (KCl)

SALSYS

ca	Input: {K_2Cl_2}			Output: {$MgCl_2$}
°C	min.		max.	
0	0.6	–	3.8	
25	0.9	–	5.5	
55	1.3	–	8.0	
83	1.8	–	10.5	
105	2.3	–	12.6	

sy	Input: {$MgCl_2$}			Output: {K_2Cl_2}
°C	min.		max.	
0	0	–	68.1	
25	0	–	71.8	
55	0	–	78.2	
83	0	–	86.0	
105	0	–	93.5	

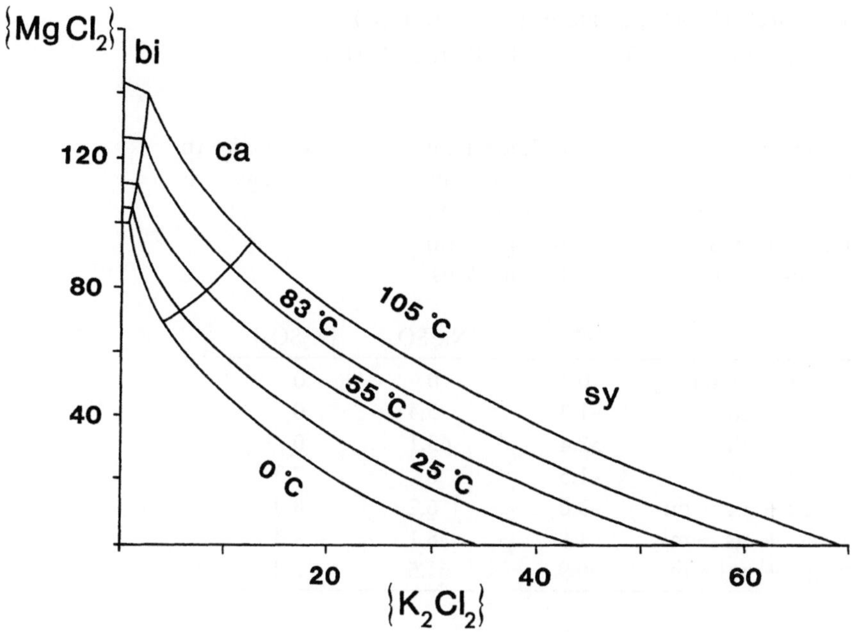
{MgCl2}
bi
ca
120
80
40
83 °C
105 °C
55 °C
sy
25 °C
0 °C
20
40
60
{K2Cl2}

III.3.4
Na_2SO_4-K_2SO_4-H_2O

ac: arcanite (K_2SO_4), gs: glaserite ($K_3Na(SO_4)_2$)
mi: mirabilite ($Na_2SO_4 \cdot 10H_2O$), th: thenardite (Na_2SO_4)

1–2–a–5:	ice	a–2: ice + mi		c–3: mi + th
2–3–c–b–a:	mi	a–5: ice + ac		c–d: gs + th
3–4–d–c:	th	a–b: ac + mi		
5–6–e–b–a:	ac	b–c: gs + mi		
b–c–d–e:	gs	b–e: ac + gs		

	°C	{Na_2SO_4}	{K_2SO_4}
1: ice + water	0	0	0
2: ice + mi	–1.3	5.4	0
3: mi + th	32.4	63.1	0
5: ice + ac	–1.5	0	7.3
a: ice + ac + mi	–3.0	6.2	8.9
b: ac + gs + mi	1.8	8.1	9.3
c: gs + mi + th	30.9	61.5	7.8

SALSYS
Input: temperature. Output: {Na_2SO_4}, {K_2SO_4}

	°C		
b–c: gs + mi	1.8	-	30.9
b–e: ac + gs	1.8	-	100.0
c–d: gs + th	30.9	-	100.0

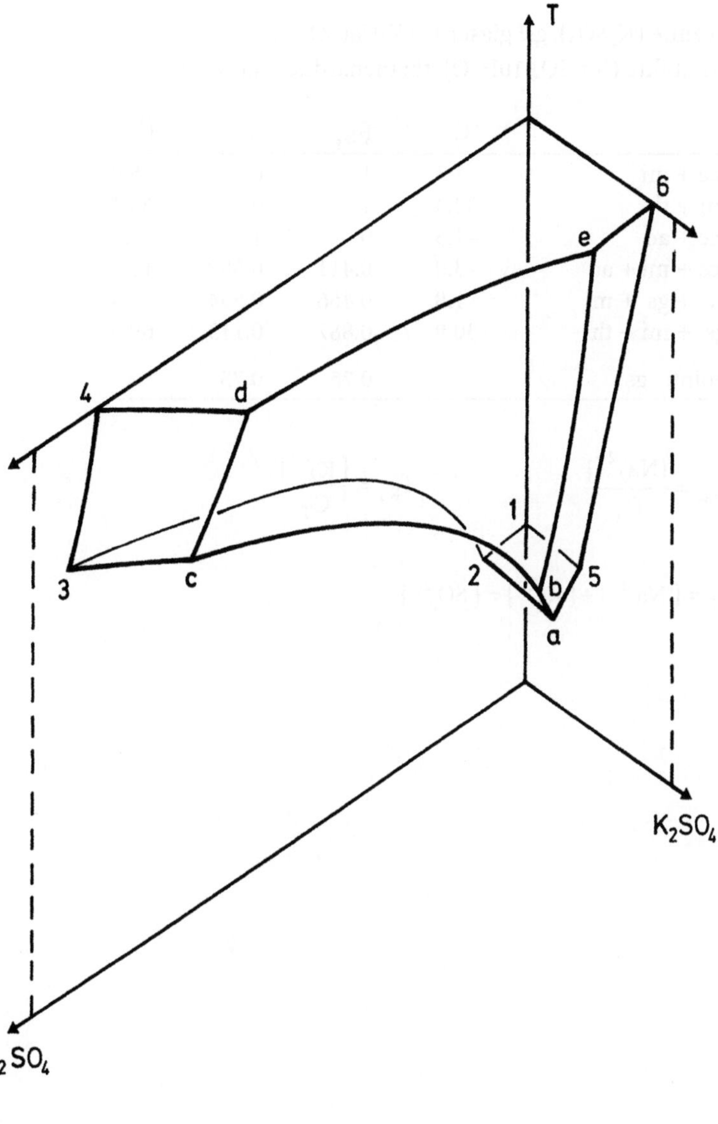
T
6
e
4
d
1
2
5
3
c
b
a
K2SO4
Na2SO4

III.3.4
Na_2SO_4-K_2SO_4-H_2O

ac: arcanite (K_2SO_4), gs: glaserite ($K_3Na(SO_4)_2$)
mi: mirabilite ($Na_2SO_4 \cdot 10H_2O$), th: thenardite (Na_2SO_4)

	°C	f_{Na_2}	f_{K_2}	C_T
2: ice + mi	−1.3	1	0	5.4
3: mi + th	32.4	1	0	63.1
5: ice + ac	−1.5	0	1	7.3
a: ice + mi + ac	−3.0	0.411	0.589	15.1
b: ac + gs + mi	1.8	0.466	0.534	17.4
c: gs + mi + th	30.9	0.887	0.113	69.3
salt point: gs		0.25	0.75	

$$f_{Na_2} = \frac{\{Na_2^{2+}\}}{C_T} \qquad f_{K_2} = \frac{\{K_2^{2+}\}}{C_T}$$

$$C_T = \{Na_2^{2+}\} + \{K_2^{2+}\} = \{SO_4^{2-}\}$$

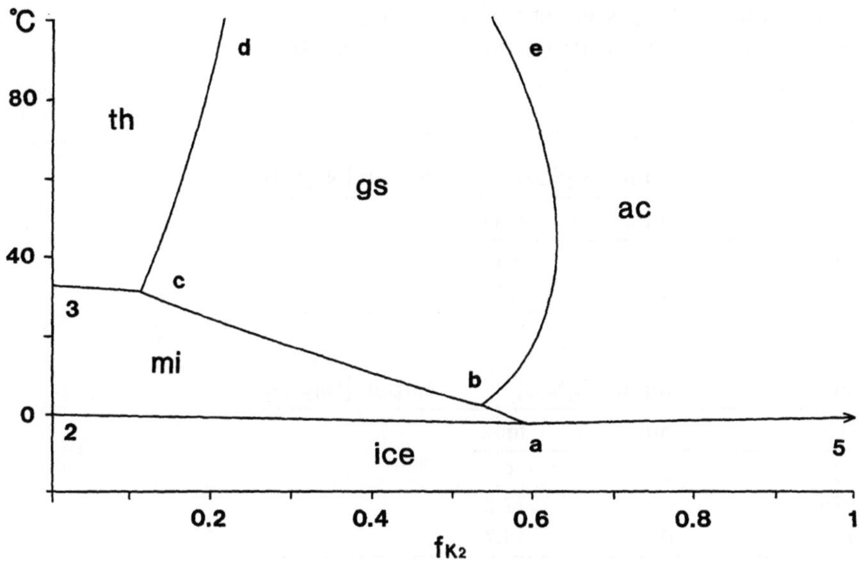

°C
80
40
0
th
gs
ac
mi
ice
d
e
c
b
a
3
2
5
0.2
0.4
0.6
0.8
1
fK2

III.3.4
Na_2SO_4-K_2SO_4-H_2O

ac: arcanite (K_2SO_4), gs: glaserite ($K_3Na(SO_4)_2$)
mi: mirabilite ($Na_2SO_4 \cdot 10H_2O$), th: thenardite (Na_2SO_4)

SALSYS

mi	Input: {K_2SO_4}			Output: {Na_2SO_4}
°C	min.		max.	
15	0	–	9.6	
25	0	–	8.7	

th	Input: {K_2SO_4}			Output: {Na_2SO_4}
°C	min.		max.	
50	0	–	9.7	
75	0	–	12.2	
100	0	–	14.7	

ac	Input: {Na_2SO_4}			Output: {K_2SO_4}
°C	min.		max.	
15	0	–	8.2	
25	0	–	8.5	
75	0	–	12.8	
100	0	–	17.3	

gs	Input: {Na_2SO_4}			Output: {K_2SO_4}
°C	min.		max.	
15	8.2	–	19.5	
25	8.5	–	42.3	
50	10.2	–	57.6	
75	12.8	–	54.2	
100	17.3	–	52.9	

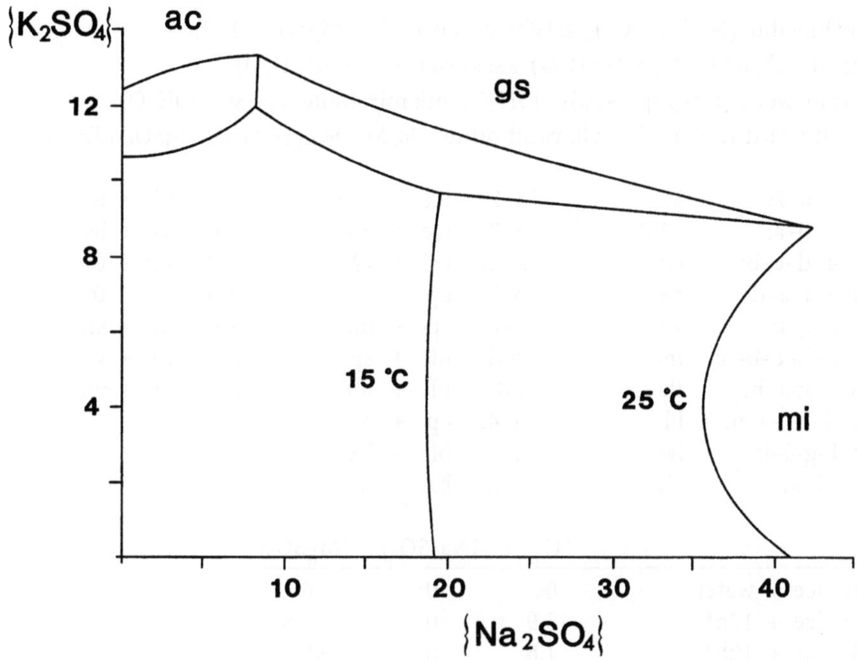

{K2SO4}
ac
gs
12
8
4
15 °C
25 °C
mi
10
20
30
40
{Na2SO4}

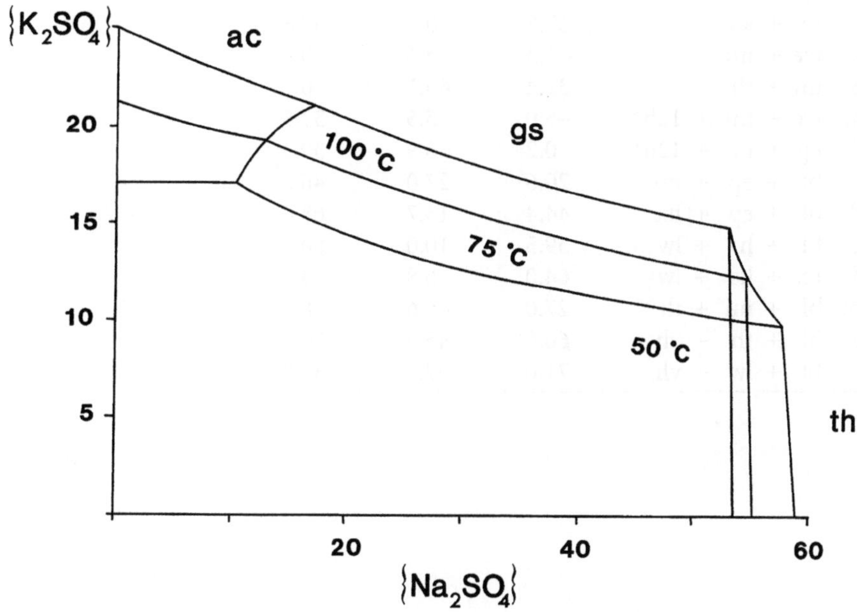

{K2SO4}
ac
gs
20
15
10
5
100 °C
75 °C
50 °C
th
20
40
60
{Na2SO4}

III.3.5
Na_2SO_4-$MgSO_4$-H_2O

bl: bloedite ($Na_2Mg(SO_4)_2 \cdot 4H_2O$), ep: epsomite ($MgSO_4 \cdot 7H_2O$)
hx: hexahydrite ($MgSO_4 \cdot 6H_2O$), ks: kieserite ($MgSO_4 \cdot H_2O$)
lw: loeweite ($Na_{12}Mg_7(SO_4)_{13} \cdot 15H_2O$), mi: mirabilite ($Na_2SO_4 \cdot 10H_2O$)
th: thenardite (Na_2SO_4), vh: vanthoffite ($Na_6Mg(SO_4)_4$), 12h*: $MgSO_4 \cdot 12H_2O$

1–2–a–7:	ice	a–2:	ice + 12h*	e–j:	bl + lw
2–3–b–a:	12h*	a–7:	ice + mi	f–g:	ks + lw
3–4–d–c–b:	ep	a–b:	mi + 12h*	h–8:	mi + th
4–5–f–e–d:	hx	b–3:	ep + 12h*	h–i:	bl + th
5–6–g–f:	ks	b–c:	ep + mi	i–j:	bl + vh
7–8–h–c–b–a:	mi	c–d:	bl + ep	i–k:	th + vh
8–9–k–i–h:	th	c–h:	bl + mi	j–l:	lw + vh
c–d–e–j–i–h:	bl	d–4:	ep + hx		
e–f–g–l–j:	lw	d–e:	bl + hx		
i–j–l–k:	vh	e–f:	hx + lw		

	°C	{Na_2SO_4}	{$MgSO_4$}
1: ice + water	0	0	0
2: ice + 12h*	−3.9	0	32.8
3: ep + 12h*	1.8	0	40.7
4: ep + hx	48.2	0	73.4
5: hx + ks	67.5	0	87.8
7: ice + mi	−1.3	5.4	0
8: mi + th	32.4	63.1	0
a: ice + mi + 12h*	−5.0	5.5	32.6
b: ep + mi + 12h*	0.2	6.9	39.3
c: bl + ep + mi	20.6	27.0	46.3
d: bl + ep + hx	44.4	13.7	68.4
e: bl + hx + lw	59.5	10.0	80.0
f: hx + ks + lw	64.0	6.8	84.0
h: bl + mi + th	27.0	45.6	28.7
i: bl + th + vh	60.0	45.5	30.6
j: bl + lw + vh	71.0	37.5	36.0

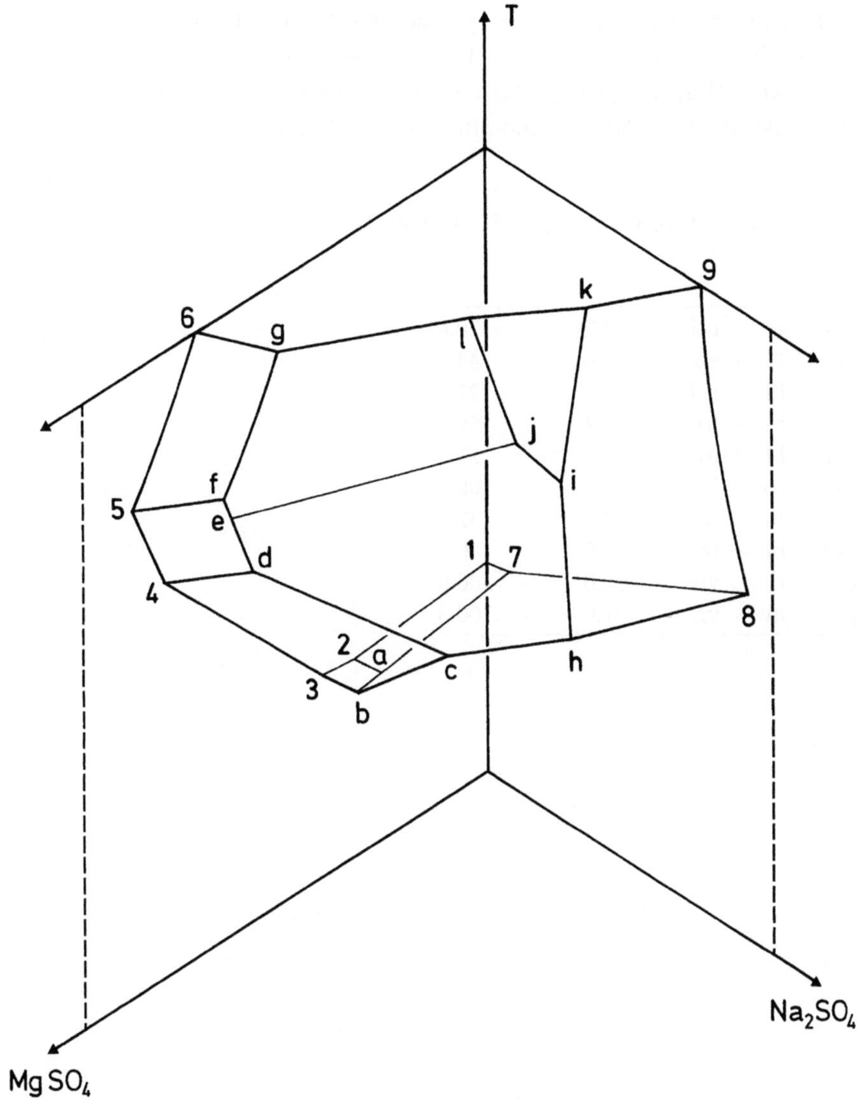
T
6
g
l
k
9
j
i
f
5
e
d
1
7
4
8
2
a
3
c
h
b
Na$_2$SO$_4$
Mg SO$_4$

III.3.5
Na_2SO_4-$MgSO_4$-H_2O

bl: bloedite ($Na_2Mg(SO_4)_2{\cdot}4H_2O$), ep: epsomite ($MgSO_4{\cdot}7H_2O$)
hx: hexahydrite ($MgSO_4{\cdot}6H_2O$), ks: kieserite ($MgSO_4{\cdot}H_2O$)
lw: loeweite ($Na_{12}Mg_7(SO_4)_{13}{\cdot}15H_2O$), mi: mirabilite ($Na_2SO_4{\cdot}10H_2O$)
th: thenardite (Na_2SO_4), vh: vanthoffite ($Na_6Mg(SO_4)_4$)

SALSYS
Input: temperature. Output: $\{Na_2SO_4\}$, $\{MgSO_4\}$

		°C		
b–c:	ep + mi	0.2	–	20.6
c–d:	bl + ep	20.6	–	44.4
c–h:	bl + mi	20.6	–	27.0
d–e:	bl + hx	44.4	–	59.5
e–j:	bl + lw	59.5	–	71.0
f–g:	ks + lw	64.0	–	100.0
h–8:	mi + th	27.0	–	32.4
h–i:	bl + th	27.0	–	60.0
i–k:	th + vh	60.0	–	100.0
j–l:	lw + vh	71.0	–	100.0

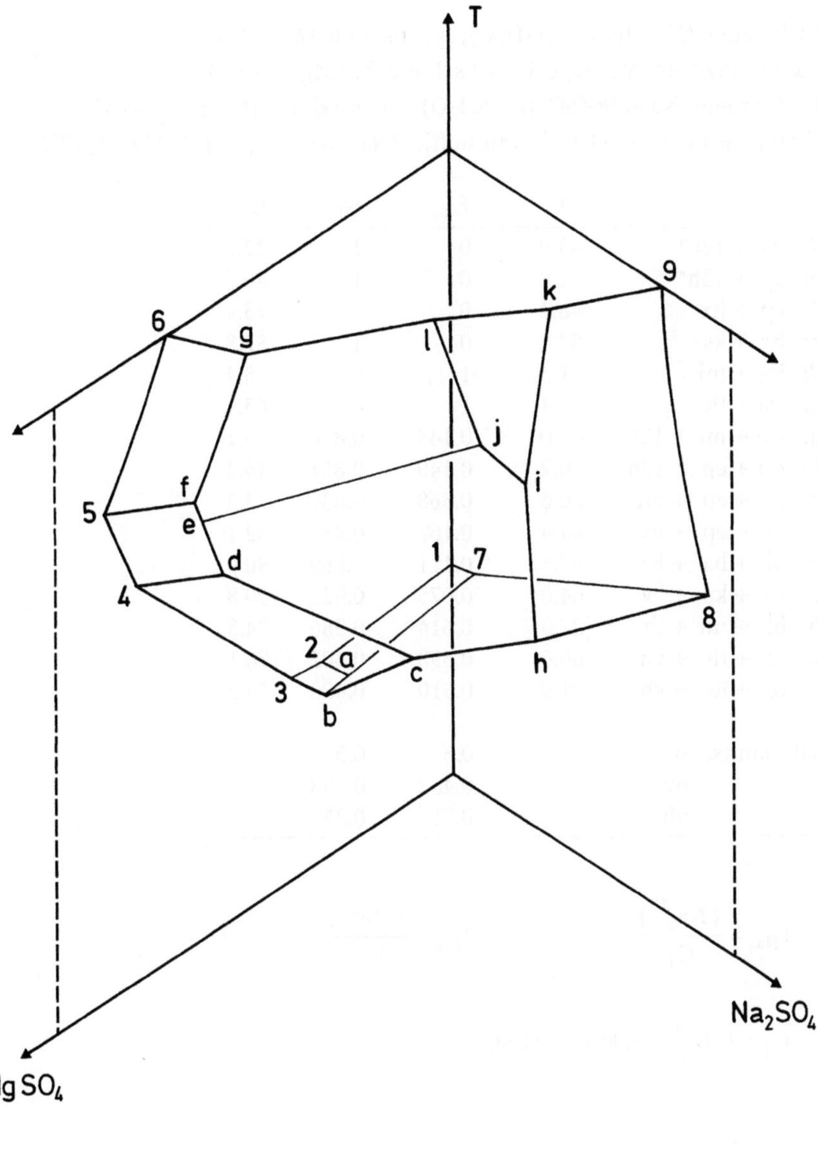
T
6
g
l
k
9
j
i
5
f
e
d
4
1
7
8
2
a
3
b
c
h
Mg SO$_4$
Na$_2$SO$_4$

III.3.5
Na_2SO_4-$MgSO_4$-H_2O

bl: bloedite ($Na_2Mg(SO_4)_2{\cdot}4H_2O$), ep: epsomite ($MgSO_4{\cdot}7H_2O$)
hx: hexahydrite ($MgSO_4{\cdot}6H_2O$), ks: kieserite ($MgSO_4{\cdot}H_2O$)
lw: loeweite ($Na_{12}Mg_7(SO_4)_{13}{\cdot}15H_2O$), mi: mirabilite ($Na_2SO_4{\cdot}10H_2O$)
th: thenardite (Na_2SO_4), vh: vanthoffite ($Na_6Mg(SO_4)_4$), 12h*: $MgSO_4{\cdot}12H_2O$

	°C	f_{Na_2}	f_{Mg}	C_T
2: ice + 12h*	−3.9	0	1	32.8
3: ep + 12h*	1.8	0	1	40.7
4: ep + hx	48.2	0	1	73.4
5: hx + ks	67.5	0	1	87.8
7: ice + mi	−1.3	1	0	5.4
8: mi + th	32.4	1	0	63.1
a: ice + mi + 12h*	−5.0	0.144	0.856	38.1
b: mi + ep + 12h*	0.2	0.149	0.851	46.2
c: bl + ep + mi	20.6	0.368	0.632	73.3
d: bl + ep + hx	44.4	0.167	0.833	82.1
e: bl + hx + lw	59.5	0.111	0.889	90.0
f: hx + ks + lw	64.0	0.075	0.925	90.8
h: bl + mi + th	27.0	0.614	0.386	74.3
i: bl + th + vh	60.0	0.598	0.402	76.1
j: bl + lw + vh	71.0	0.510	0.490	73.5
salt points: bl		0.5	0.5	
lw		0.462	0.538	
vh		0.75	0.25	

$$f_{Na_2} = \frac{\{Na_2^{2+}\}}{C_T} \qquad f_{Mg} = \frac{\{Mg^{2+}\}}{C_T}$$

$$C_T = \{Na_2^{2+}\} + \{Mg^{2+}\} = \{SO_4^{2-}\}$$

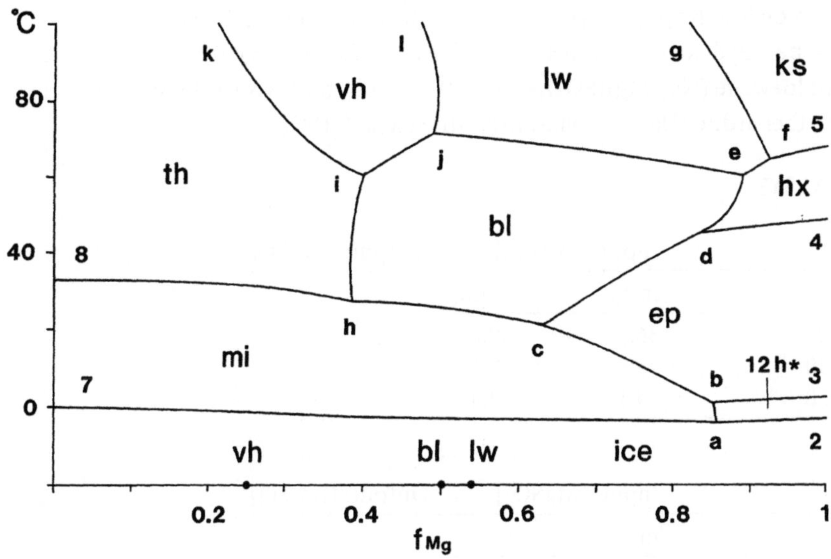
°C
80
40
0
0.2
0.4
0.6
0.8
1
fMg
k
l
g
ks
vh
lw
5
f
e
j
th
i
hx
bl
40
8
d
4
ep
h
c
mi
7
12h*
b
3
a
2
vh
bl
lw
ice

III.3.5
Na_2SO_4-$MgSO_4$-H_2O

bl: bloedite ($Na_2Mg(SO_4)_2{\cdot}4H_2O$), ep: epsomite ($MgSO_4{\cdot}7H_2O$)
hx: hexahydrite ($MgSO_4{\cdot}6H_2O$), ks: kieserite ($MgSO_4{\cdot}H_2O$)
lw: loeweite ($Na_{12}Mg_7(SO_4)_{13}{\cdot}15H_2O$), mi: mirabilite ($Na_2SO_4{\cdot}10H_2O$)
th: thenardite (Na_2SO_4), vh: vanthoffite ($Na_6Mg(SO_4)_4$)

SALSYS

bl	Input: {$MgSO_4$}			Output: {Na_2SO_4}
°C	min.		max.	
25	36.1	–	49.6	
30	28.5	–	53.7	
50	29.1	–	73.4	

lw	Input: {$MgSO_4$}			Output: {Na_2SO_4}
°C	min.		max.	
75	36.0	–	76.9	
100	32.6	–	62.8	

mi	Input: {$MgSO_4$}			Output: {Na_2SO_4}
°C	min.		max.	
10	0	–	42.8	
15	0	–	44.5	
25	0	–	36.1	
30	0	–	19.7	

th	Input: {$MgSO_4$}			Output: {Na_2SO_4}
°C	min.		max.	
30	19.7	–	28.5	
50	0	–	29.1	
75	0	–	20.7	
100	0	–	13.3	

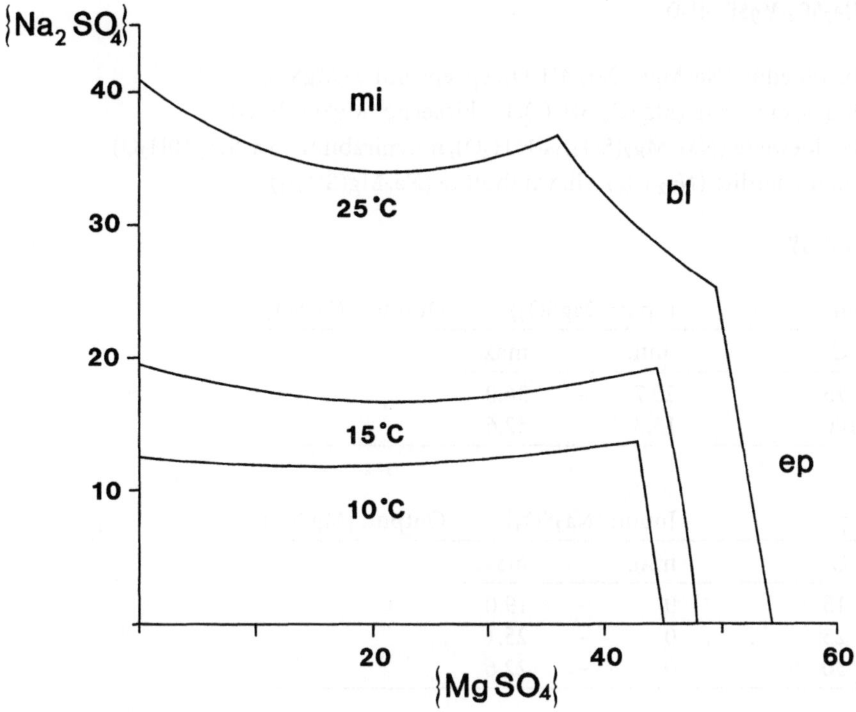
{Na2 SO4}
40
30
20
10
mi
25 °C
bl
15 °C
10 °C
ep
20
40
60
{Mg SO4}

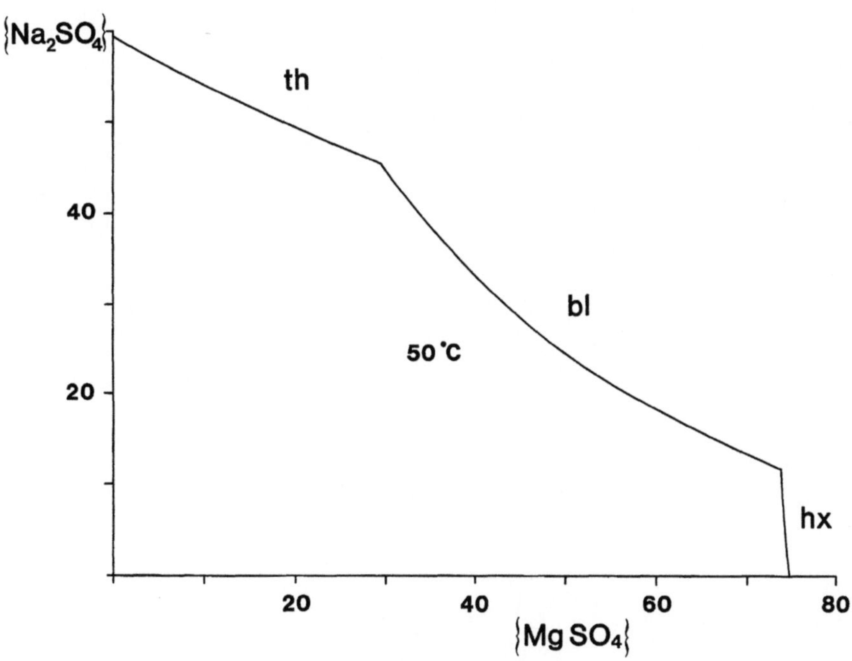
{Na2SO4}
40
20
th
bl
50 °C
hx
20
40
60
80
{Mg SO4}

III.3.5
Na_2SO_4-$MgSO_4$-H_2O

bl: bloedite ($Na_2Mg(SO_4)_2\cdot4H_2O$), ep: epsomite ($MgSO_4\cdot7H_2O$)
hx: hexahydrite ($MgSO_4\cdot6H_2O$), ks: kieserite ($MgSO_4\cdot H_2O$)
lw: loeweite ($Na_{12}Mg_7(SO_4)_{13}\cdot15H_2O$), mi: mirabilite ($Na_2SO_4\cdot10H_2O$)
th: thenardite (Na_2SO_4), vh: vanthoffite ($Na_6Mg(SO_4)_4$)

SALSYS

vh	Input: {$MgSO_4$}			Output: {Na_2SO_4}
°C	min.		max.	
75	20.7	–	36.0	
100	13.3	–	32.6	

ep	Input: {Na_2SO_4}			Output: {$MgSO_4$}
°C	min.		max.	
15	0	–	19.0	
25	0	–	25.1	
30	0	–	22.6	

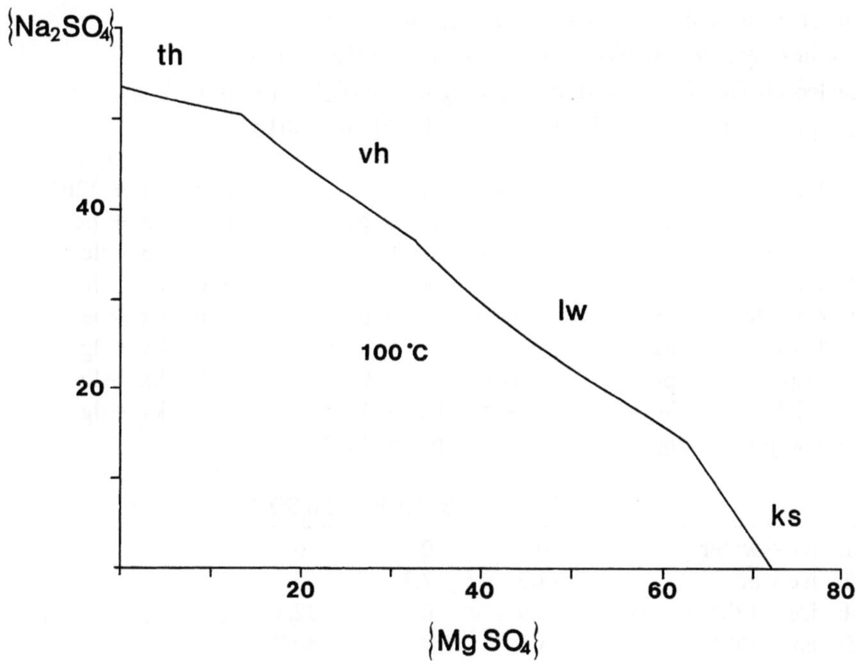
{Na2SO4}
th
vh
40
lw
100 °C
20
ks
20
40
60
80
{Mg SO4}

III.3.6
K_2SO_4-$MgSO_4$-H_2O

ac: arcanite (K_2SO_4), ep: epsomite ($MgSO_4 \cdot 7H_2O$)
hx: hexahydrite ($MgSO_4 \cdot 6H_2O$), ks: kieserite ($MgSO_4 \cdot H_2O$)
le: leonite ($K_2Mg(SO_4)_2 \cdot 4H_2O$), lg: langbeinite ($K_2Mg_2(SO_4)_3$)
pc: picromerite ($K_2Mg(SO_4)_2 \cdot 6H_2O$), 12h*: $MgSO_4 \cdot 12H_2O$

1–2–a–e–4:	ice	a–2:	ice + ac	f–5:	ep + 12h*
2–3–d–c–b–a:	ac	a–b:	ac + pc	f–g:	ep + pc
4–5–f–e:	12h*	a–e:	ice + pc	g–h:	ep + le
5–6–h–g–f:	ep	b–c:	ac + le	h–6:	ep + hx
6–7–j–i–h:	hx	b–g:	le + pc	h–i:	hx + le
7–8–k–j:	ks	c–d:	ac + lg	i–j:	hx + lg
a–b–g–f–e:	pc	c–i:	le + lg	j–7:	hx + ks
b–c–i–h–g:	le	e–4:	ice + 12h*	j–k:	ks + lg
c–d–k–j–i:	lg	e–f:	pc + 12h*		

	°C	{K_2SO_4}	{$MgSO_4$}
1: ice + water	0	0	0
2: ice + ac	-1.5	7.3	0
4: ice + 12h*	-3.9	0	32.8
5: ep + 12h*	1.8	0	40.7
6: ep + hx	48.2	0	73.4
7: hx + ks	67.5	0	87.8
a: ice + ac + pc	-3.0	9.0	15.4
b: ac + le + pc	47.5	18.8	36.2
c: ac + le + lg	89.0	27.0	39.7
e: ice + pc + 12h*	-5.2	4.5	33.7
f: ep + pc + 12h*	1.0	4.2	41.3
g: ep + le + pc	41.0	7.3	72.0
h: ep + hx + le	47.2	7.1	75.3
i: hx + le + lg	61.0	7.3	85.8
j: hx + ks + lg	66.5	7.1	88.8

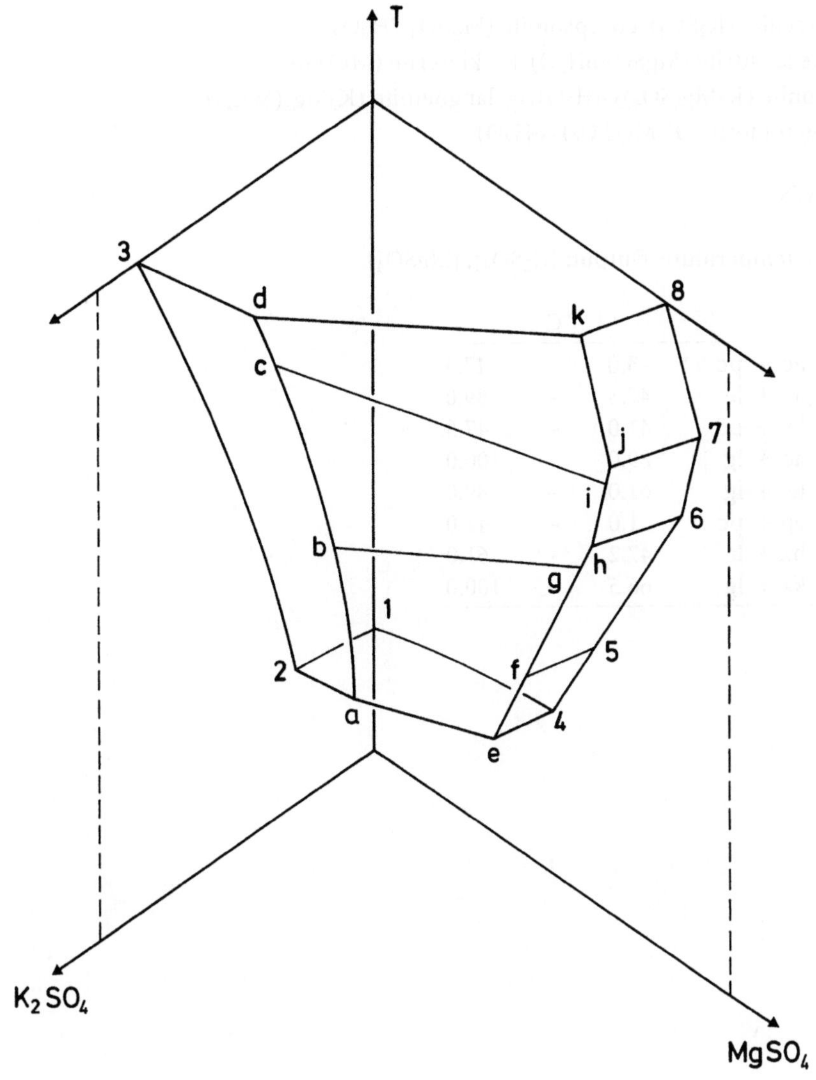
T
3
d
k
8
c
7
j
i
6
b
h
g
1
5
2
f
a
4
e
K_2SO_4
$MgSO_4$

III.3.6
K_2SO_4-$MgSO_4$-H_2O

ac: arcanite (K_2SO_4), ep: epsomite ($MgSO_4{\cdot}7H_2O$)
hx: hexahydrite ($MgSO_4{\cdot}6H_2O$), ks: kieserite ($MgSO_4{\cdot}H_2O$)
le: leonite ($K_2Mg(SO_4)_2{\cdot}4H_2O$), lg: langbeinite ($K_2Mg_2(SO_4)_3$)
pc: picromerite ($K_2Mg(SO_4)_2{\cdot}6H_2O$)

SALSYS

Input: temperature. Output: {K_2SO_4}, {$MgSO_4$}

		°C	
a–b: ac + pc	−3.0	–	47.5
b–c: ac + le	47.5	–	89.0
b–g: le + pc	41.0	–	47.5
c–d: ac + lg	89.0	–	100.0
c–i: le + lg	61.0	–	89.0
f–g: ep + pc	1.0	–	41.0
h–i: hx + le	47.2	–	61.0
j–k: ks + lg	66.5	–	100.0

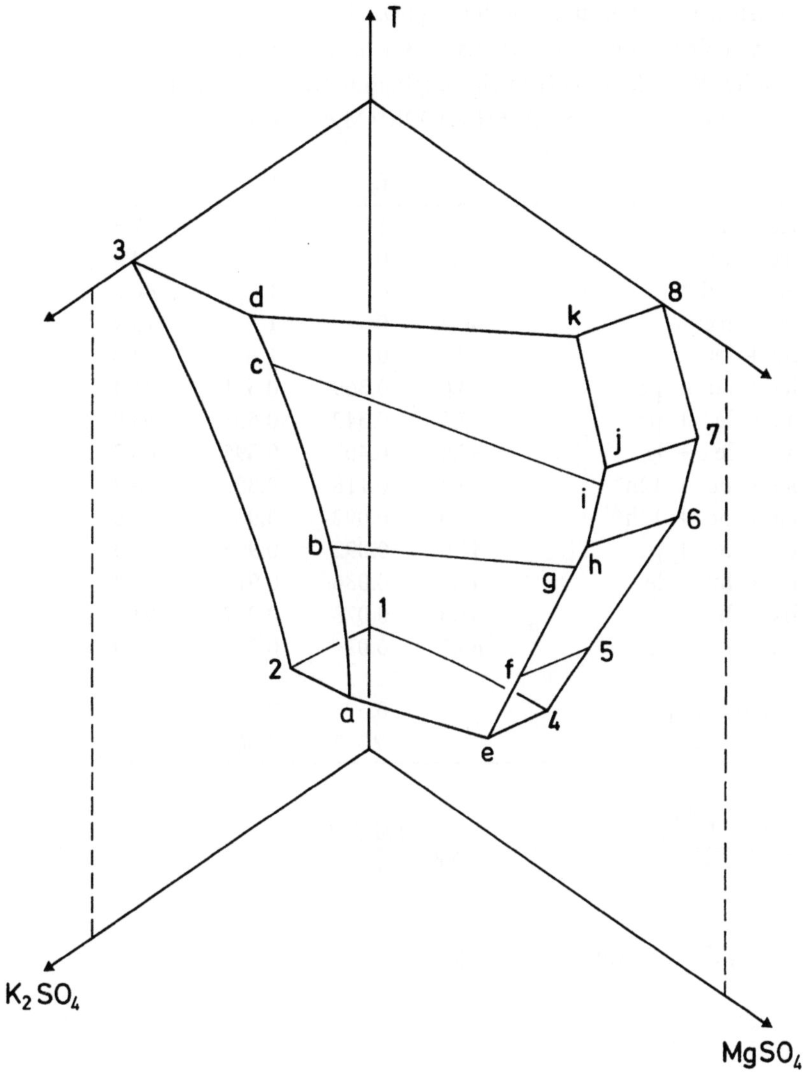
T
3
d
k
8
c
j
7
i
6
b
h
g
1
5
2
f
a
4
e
K_2SO_4
$MgSO_4$

III.3.6
K_2SO_4-$MgSO_4$-H_2O

ac: arcanite (K_2SO_4), ep: epsomite ($MgSO_4 \cdot 7H_2O$)
hx: hexahydrite ($MgSO_4 \cdot 6H_2O$), ks: kieserite ($MgSO_4 \cdot H_2O$)
le: leonite ($K_2Mg(SO_4)_2 \cdot 4H_2O$), lg: langbeinite ($K_2Mg_2(SO_4)_3$)
pc: picromerite ($K_2Mg(SO_4)_2 \cdot 6H_2O$), 12h*: $MgSO_4 \cdot 12H_2O$

	°C	f_{K_2}	f_{Mg}	C_T
2: ice + ac	−1.5	1	0	7.3
4: ice + 12h*	−3.9	0	1	32.8
5: ep + 12h*	1.8	0	1	40.7
6: ep + hx	48.2	0	1	73.4
7: hx + ks	67.5	0	1	87.8
a: ice + ac + pc	−3.0	0.369	0.631	24.4
b: ac + le + pc	47.5	0.342	0.658	55.0
c: ac + le + lg	89.0	0.405	0.595	66.7
e: ice + pc + 12h*	−5.2	0.118	0.882	38.2
f: ep + pc + 12h*	1.0	0.092	0.908	45.5
g: ep + le + pc	41.0	0.092	0.908	79.3
h: ep + hx + le	47.2	0.086	0.914	82.4
i: hx + le + lg	61.0	0.078	0.922	93.1
j: hx + ks + lg	66.5	0.074	0.926	95.9
salt points: le, pc		0.5	0.5	
lg		0.333	0.667	

$$f_{K_2} = \frac{\{K_2^{2+}\}}{C_T} \qquad f_{Mg} = \frac{\{Mg^{2+}\}}{C_T}$$

$$C_T = \{K_2^{2+}\} + \{Mg^{2+}\} = \{SO_4^{2-}\}$$

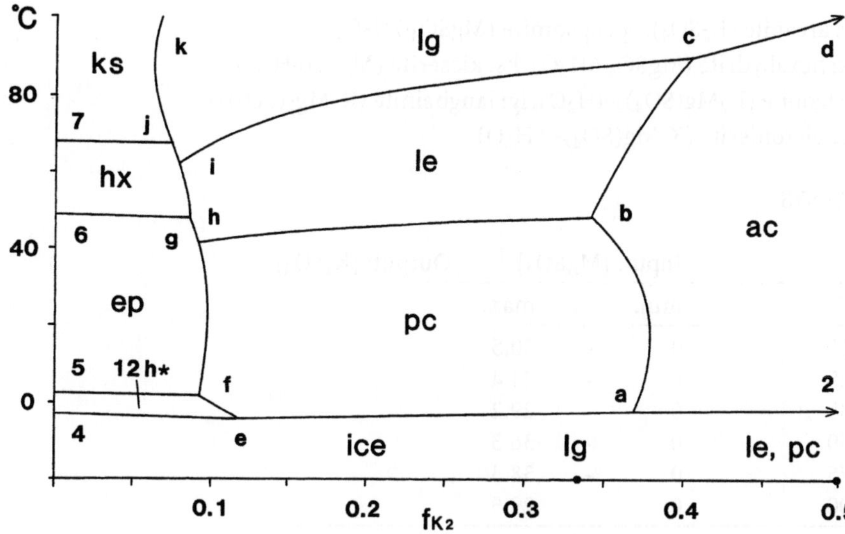
°C
80
40
0
ks
k
lg
c
d
7
j
hx
i
le
h
b
ac
6
g
ep
pc
5
12
h*
f
2
a
4
e
ice
lg
le, pc
0.1
0.2
fK2
0.3
0.4
0.5

III.3.6
K_2SO_4-$MgSO_4$-H_2O

ac: arcanite (K_2SO_4), ep: epsomite ($MgSO_4 \cdot 7H_2O$)
hx: hexahydrite ($MgSO_4 \cdot 6H_2O$), ks: kieserite ($MgSO_4 \cdot H_2O$)
le: leonite ($K_2Mg(SO_4)_2 \cdot 4H_2O$), lg: langbeinite ($K_2Mg_2(SO_4)_3$)
pc: picromerite ($K_2Mg(SO_4)_2 \cdot 6H_2O$)

SALSYS

ac	Input: {$MgSO_4$}			Output: {K_2SO_4}
°C	min.		max.	
15	0	–	20.5	
25	0	–	24.4	
35	0	–	29.2	
50	0	–	36.3	
75	0	–	38.3	
100	0	–	29.5	

le	Input: {$MgSO_4$}			Output: {K_2SO_4}
°C	min.		max.	
50	36.3	–	76.6	
75	38.3	–	70.2	

lg	Input: {$MgSO_4$}			Output: {K_2SO_4}
°C	min.		max.	
75	70.2	–	84.1	
100	29.5	–	69.5	

pc	Input: {$MgSO_4$}			Output: {K_2SO_4}
°C	min.		max.	
15	20.5	–	49.3	
25	24.4	–	56.8	
35	29.2	–	65.9	

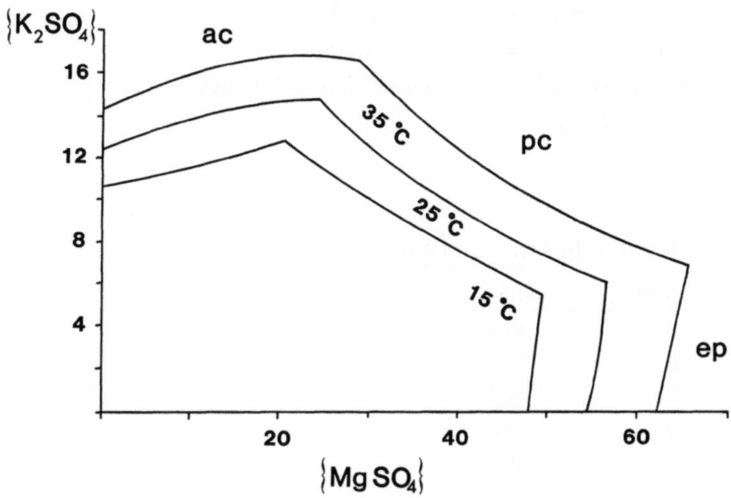

{K2SO4}
ac
16
12
8
4
35 °C
pc
25 °C
15 °C
ep
20
40
60
{Mg SO4}

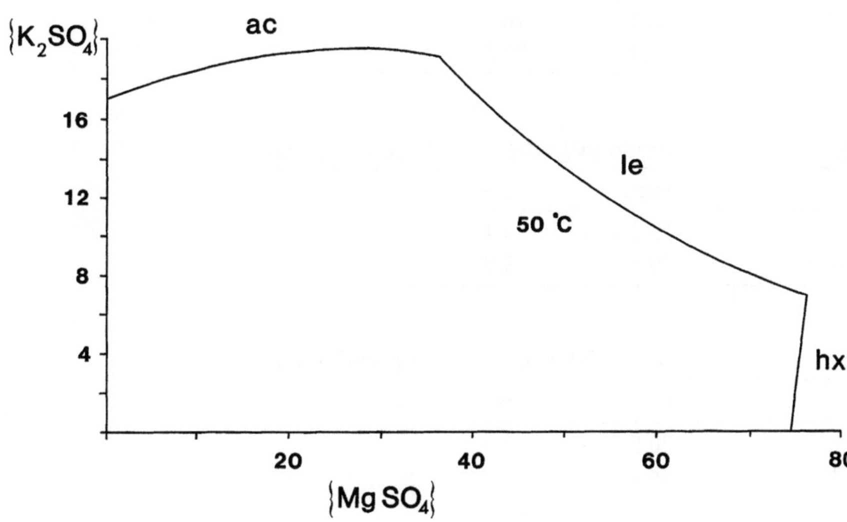

{K2SO4}
ac
16
12
8
4
le
50 °C
hx
20
40
60
80
{Mg SO4}

III.3.6
K_2SO_4-$MgSO_4$-H_2O

ac: arcanite (K_2SO_4), ep: epsomite ($MgSO_4 \cdot 7H_2O$)
hx: hexahydrite ($MgSO_4 \cdot 6H_2O$), ks: kieserite ($MgSO_4 \cdot H_2O$)
le: leonite ($K_2Mg(SO_4)_2 \cdot 4H_2O$), lg: langbeinite ($K_2Mg_2(SO_4)_3$)
pc: picromerite ($K_2Mg(SO_4)_2 \cdot 6H_2O$)

SALSYS

ac	Input: {$MgSO_4$}			Output: {K_2SO_4}
°C	min.		max.	
15	0	–	20.5	
25	0	–	24.4	
35	0	–	29.2	
50	0	–	36.3	
75	0	–	38.3	
100	0	–	29.5	

le	Input: {$MgSO_4$}			Output: {K_2SO_4}
°C	min.		max.	
50	36.3	–	76.6	
75	38.3	–	70.2	

lg	Input: {$MgSO_4$}			Output: {K_2SO_4}
°C	min.		max.	
75	70.2	–	84.1	
100	29.5	–	69.5	

pc	Input: {$MgSO_4$}			Output: {K_2SO_4}
°C	min.		max.	
15	20.5	–	49.3	
25	24.4	–	56.8	
35	29.2	–	65.9	

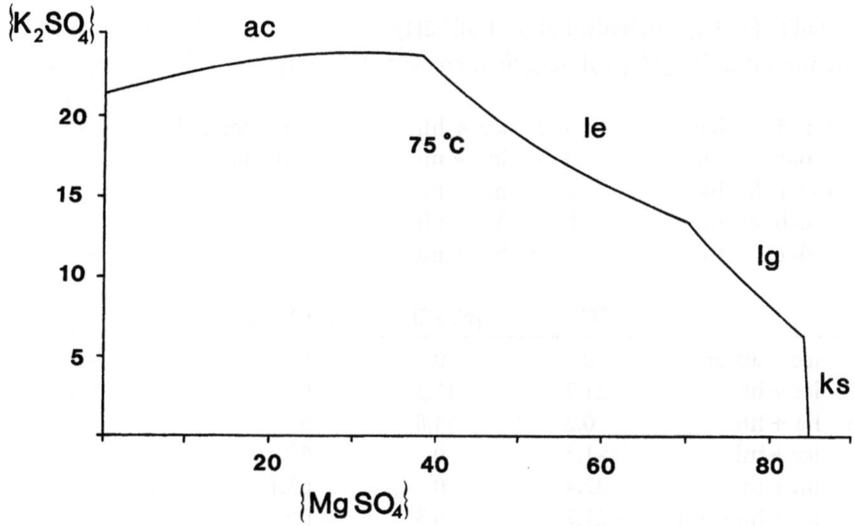
{K₂SO₄}
ac
20
15
10
5
75 °C
le
lg
ks
20
40
60
80
{Mg SO₄}

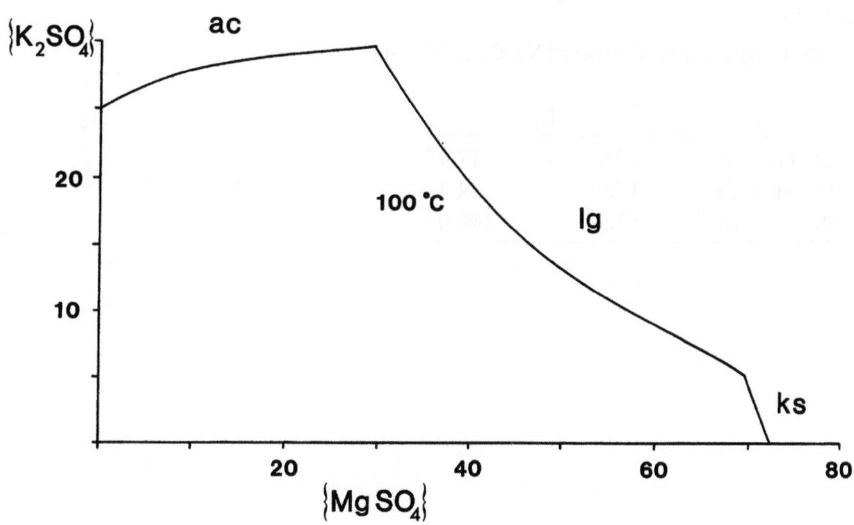
{K₂SO₄}
ac
20
10
100 °C
lg
ks
20
40
60
80
{Mg SO₄}

III.3.7
Na_2Cl_2-Na_2SO_4-H_2O

ha: halite (NaCl), hh: hydrohalite ($NaCl{\cdot}2H_2O$)
mi: mirabilite ($Na_2SO_4{\cdot}10H_2O$), th: thenardite (Na_2SO_4)

1–2–a–5:	ice	a–2:	ice + hh	c–6:	mi + th
2–3–b–a:	hh	a–5:	ice + mi	c–d:	ha + th
3–4–d–c–b:	ha	a–b:	hh + mi		
5–6–c–b–a:	mi	b–3:	ha + hh		
6–7–d–c:	th	b–c:	ha + mi		

		°C	$\{Na_2Cl_2\}$	$\{Na_2SO_4\}$
1:	ice + water	0	0	0
2:	ice + hh	–21.2	47.3	0
3:	ha + hh	0.2	54.8	0
5:	ice + mi	–1.3	0	5.4
6:	mi + th	32.4	0	63.1
a:	ice + hh + mi	–21.7	46.3	0.3
b:	ha + hh + mi	–2.9	54.0	2.0
c:	ha + mi + th	17.9	49.0	13.8

SALSYS
Input: temperature. Output: $\{Na_2Cl_2\}$, $\{Na_2SO_4\}$

		°C		
b–c:	ha + mi	-2.9	-	17.9
c–6:	mi + th	17.9	-	32.4
c–d:	ha + th	17.9	-	100.0

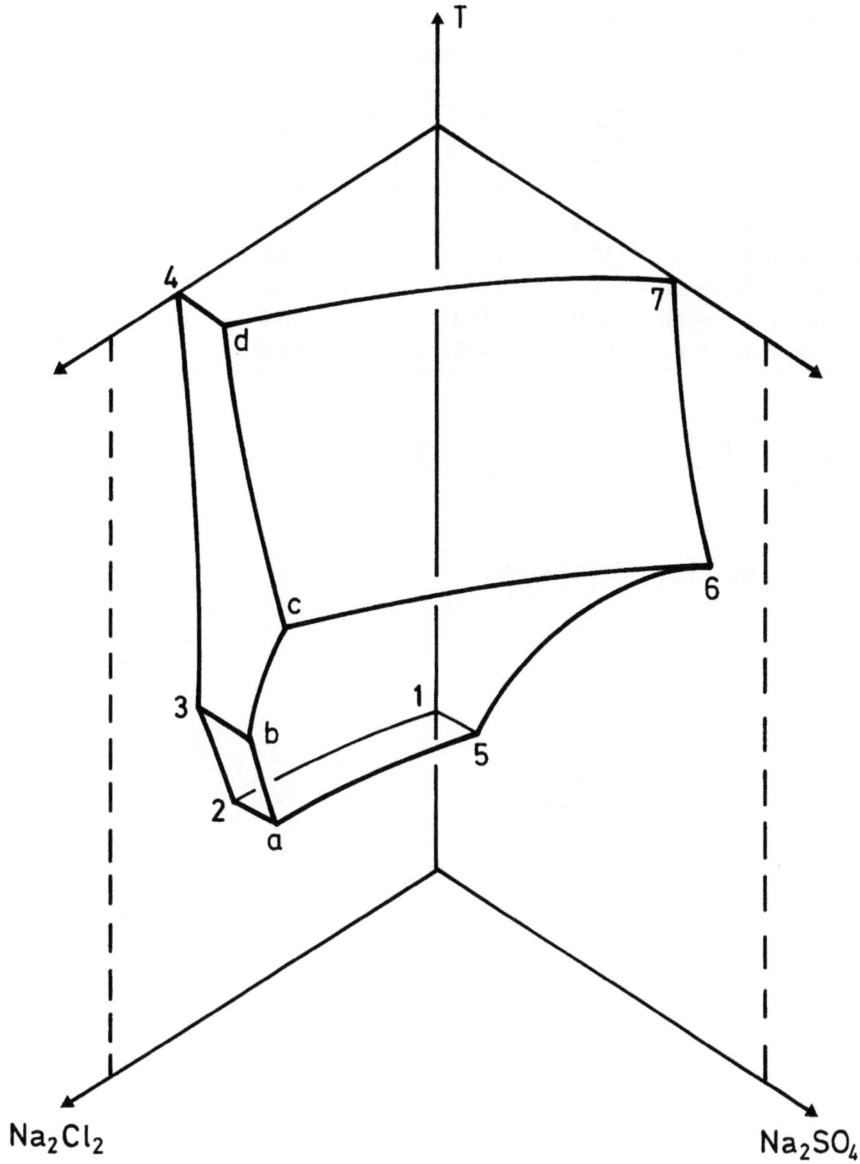
T
4
7
d
6
c
1
3
b
5
2
a
Na_2Cl_2
Na_2SO_4

III.3.7
Na_2Cl_2-Na_2SO_4-H_2O

ha: halite (NaCl), hh: hydrohalite ($NaCl \cdot 2H_2O$)
mi: mirabilite ($Na_2SO_4 \cdot 10H_2O$), th: thenardite (Na_2SO_4)

	°C	f_{Cl_2}	f_{SO_4}	C_T
2: ice + hh	−21.2	1	0	47.3
3: hh + ha	0.2	1	0	54.8
5: ice + mi	−1.3	0	1	5.4
6: mi + th	32.4	0	1	63.1
a: ice + hh + mi	−21.7	0.994	0.006	46.6
b: hh + ha + mi	−2.9	0.964	0.036	56.0
c: ha + mi + th	17.9	0.780	0.220	62.8

$$f_{Cl_2} = \frac{\{Cl_2^{2-}\}}{C_T} \qquad f_{SO_4} = \frac{\{SO_4^{2-}\}}{C_T}$$

$$C_T = \{Na_2^{2+}\} = \{Cl_2^{2-}\} + \{SO_4^{2-}\}$$

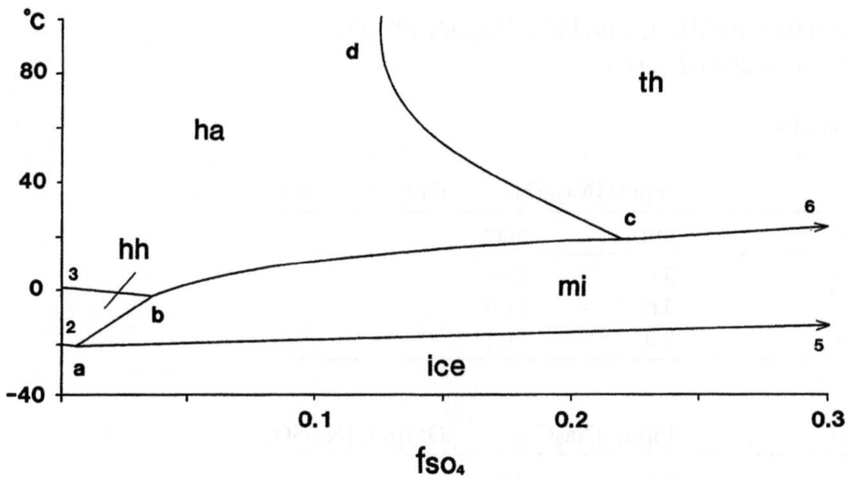
°C
80
40
0
-40
d
th
ha
hh
3
2
a
b
c
6
mi
5
ice
0.1
0.2
0.3
fso₄

III.3.7
Na_2Cl_2-Na_2SO_4-H_2O

ha: halite (NaCl), mi: mirabilite ($Na_2SO_4 \cdot 10H_2O$)
th: thenardite (Na_2SO_4)

SALSYS

ha	Input: {Na_2SO_4}			Output: {Na_2Cl_2}
°C	min.		max.	
25	5.0	–	12.6	
30	5.0	–	11.9	
35	5.0	–	11.2	

mi	Input: {Na_2Cl_2}			Output: {Na_2SO_4}
°C	min.		max.	
0	10.0	–	54.0	
10	10.0	–	52.3	
15	10.0	–	50.4	
17.9	10.0	–	49.0	
20	9.3	–	44.7	
25	10.0	–	30.6	

th	Input: {Na_2Cl_2}			Output: {Na_2SO_4}
°C	min.		max.	
25	30.6	–	49.9	
30	11.2	–	50.5	
35	4.9	–	51.0	
50	2.5	–	52.6	
75	7.5	–	55.0	
100	6.2	–	57.1	

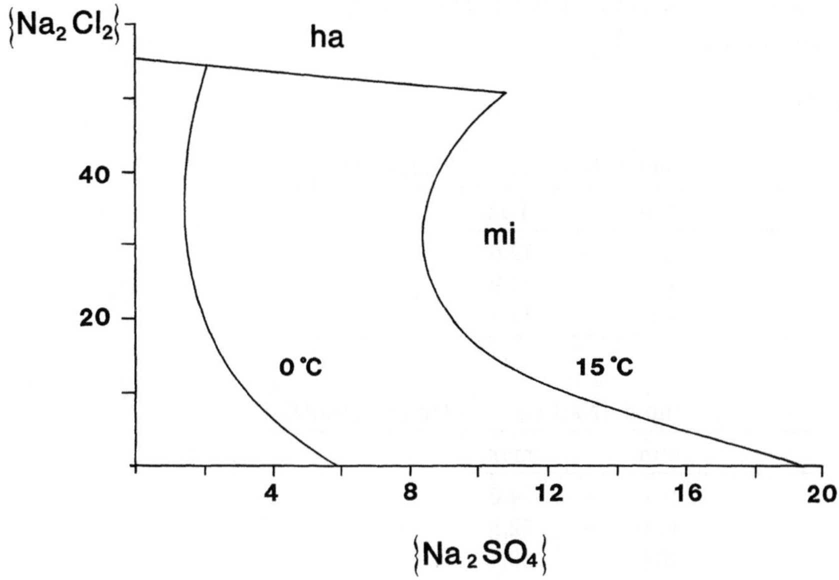
{Na2Cl2}
ha
40
mi
20
0 °C
15 °C
4
8
12
16
20
{Na2SO4}

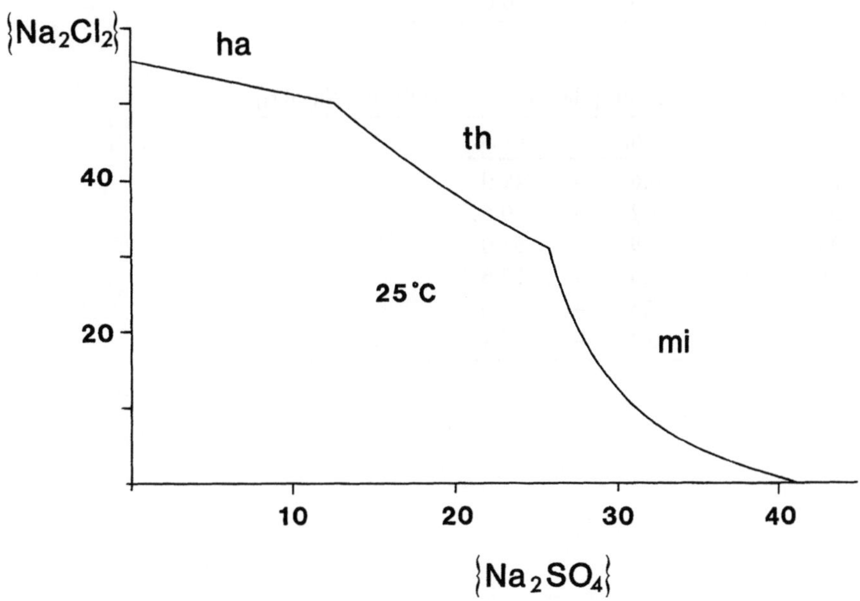
{Na2Cl2}
ha
th
40
25 °C
20
mi
10
20
30
40
{Na2SO4}

III.3.7
Na_2Cl_2-Na_2SO_4-H_2O

ha: halite (NaCl), mi: mirabilite ($Na_2SO_4 \cdot 10H_2O$)
th: thenardite (Na_2SO_4)

SALSYS

ha	Input: {Na_2SO_4}			Output: {Na_2Cl_2}
°C	min.		max.	
25	5.0	–	12.6	
30	5.0	–	11.9	
35	5.0	–	11.2	

mi	Input: {Na_2Cl_2}			Output: {Na_2SO_4}
°C	min.		max.	
0	10.0	–	54.0	
10	10.0	–	52.3	
15	10.0	–	50.4	
17.9	10.0	–	49.0	
20	9.3	–	44.7	
25	10.0	–	30.6	

th	Input: {Na_2Cl_2}			Output: {Na_2SO_4}
°C	min.		max.	
25	30.6	–	49.9	
30	11.2	–	50.5	
35	4.9	–	51.0	
50	2.5	–	52.6	
75	7.5	–	55.0	
100	6.2	–	57.1	

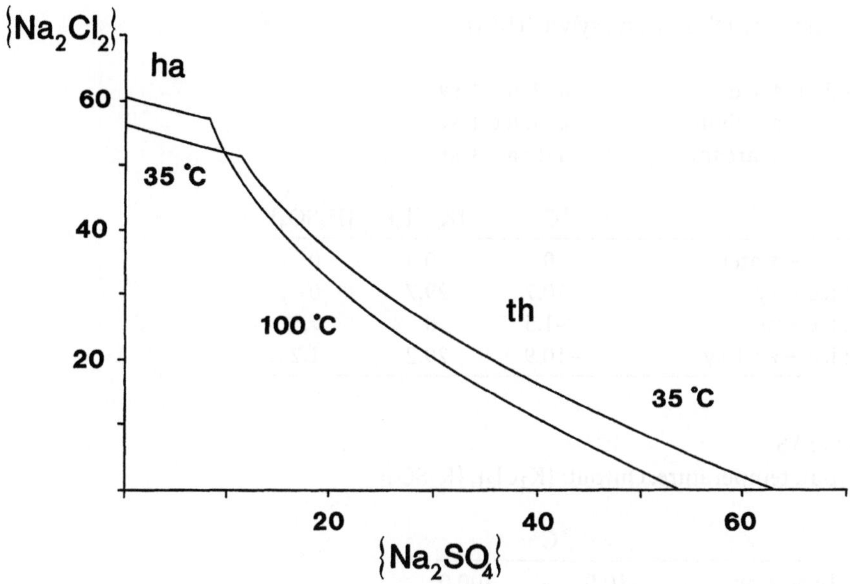
{Na2Cl2}
ha
60
35 °C
40
100 °C
th
20
35 °C
20
40
60
{Na2SO4}

III.3.8
K_2Cl_2-K_2SO_4-H_2O

ac: arcanite (K_2SO_4), sy: sylvite (KCl)

1–2–a–4: ice	a–2: ice + sy
2–3–b–a: sylvite	a–4: ice + ac
4–5–b–a: arcanite	a–b: ac + sy

	°C	$\{K_2Cl_2\}$	$\{K_2SO_4\}$
1: ice + water	0	0	0
2: ice + sy	−10.7	29.7	0
4: ice + ac	−1.5	0	7.3
a: ice + ac + sy	−10.9	29.2	1.2

SALSYS
Input: temperature. Output: $\{K_2Cl_2\}$, $\{K_2SO_4\}$

	°C		
a–b: ac + sy	−10.9	-	100.0

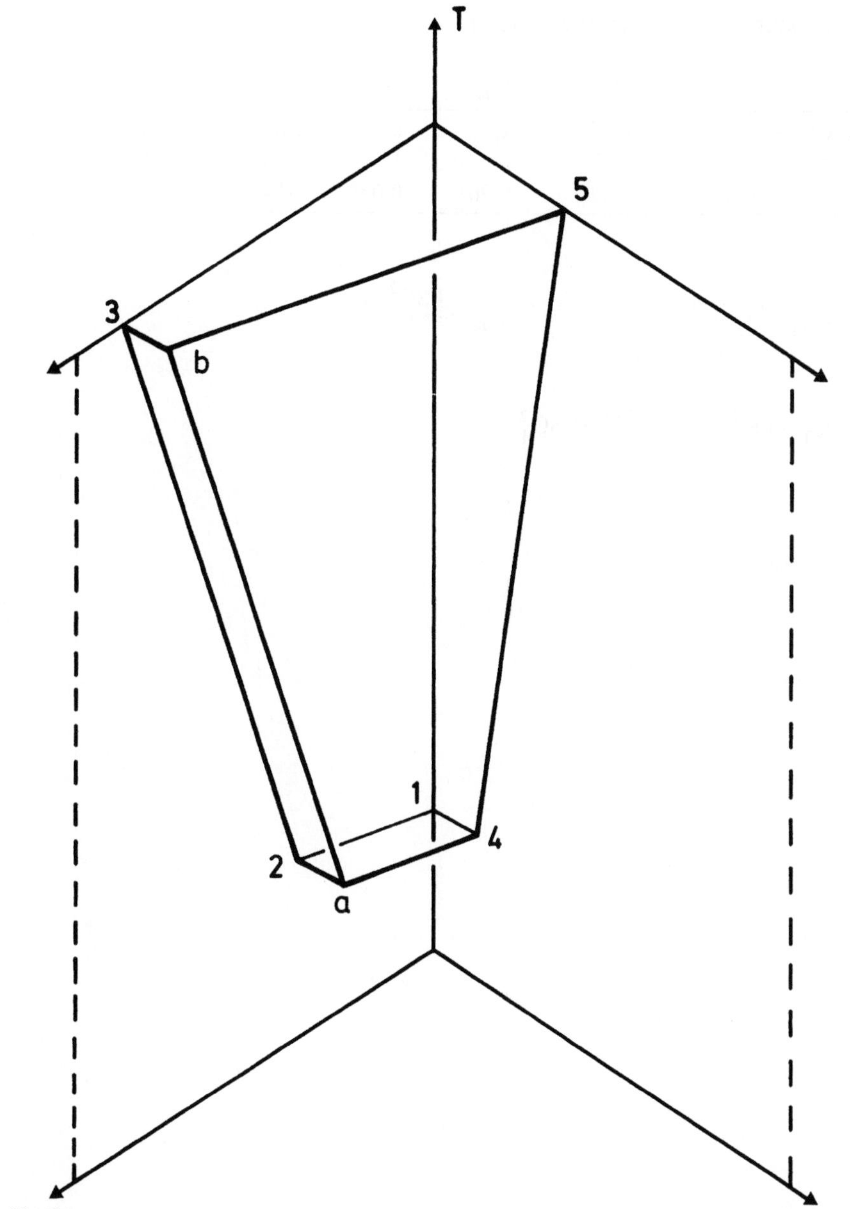
T
5
3
b
1
4
2
a
K_2Cl_2
K_2SO_4

III.3.8
K_2Cl_2-K_2SO_4-H_2O

ac: arcanite (K_2SO_4), sy: sylvite (KCl)

	°C	f_{Cl_2}	f_{SO_4}	C_T
2: ice + sy	−10.7	1	0	29.7
4: ice + ac	−1.5	0	1	7.3
a: ice + sy + ac	−10.9	0.961	0.039	30.4

$$f_{Cl_2} = \frac{\{Cl_2^{2-}\}}{C_T} \qquad f_{SO_4} = \frac{\{SO_4^{2-}\}}{C_T}$$

$$C_T = \{K_2^{2+}\} = \{Cl_2^{2-}\} + \{SO_4^{2-}\}$$

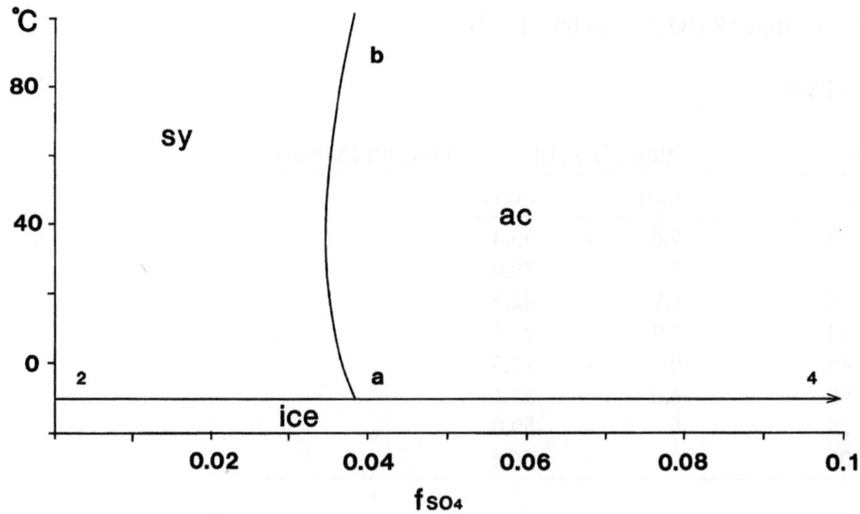
°C
80
40
0
sy
ac
b
a
2
4
ice
0.02
0.04
0.06
0.08
0.1
fSO4

III.3.8
K_2Cl_2-K_2SO_4-H_2O

ac: arcanite (K_2SO_4), sy: sylvite (KCl)

SALSYS

ac	Input: $\{K_2Cl_2\}$			Output: $\{K_2SO_4\}$
°C	min.		max.	
0	7.0	–	33.4	
15	5.3	–	39.0	
25	6.9	–	42.6	
30	3.9	–	44.3	
40	0	–	12.3	
50	4.0	–	51.1	
75	0	–	59.0	
100	0	–	66.4	

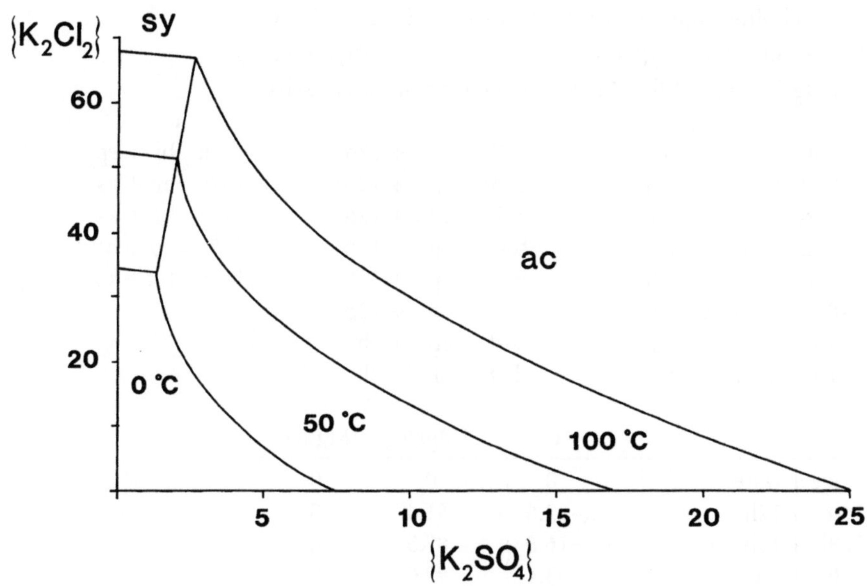
{K2Cl2}
sy
60
40
20
ac
0 °C
50 °C
100 °C
5
10
15
20
25
{K2SO4}

III.3.9
$MgCl_2$-$MgSO_4$-H_2O

bi: bischofite ($MgCl_2 \cdot 6H_2O$), ep: epsomite ($MgSO_4 \cdot 7H_2O$)
hx: hexahydrite ($MgSO_4 \cdot 6H_2O$), ks: kieserite ($MgSO_4 \cdot H_2O$)
8h: $MgCl_2 \cdot 8H_2O$, 12h: $MgCl_2 \cdot 12H_2O$, 12h*: $MgSO_4 \cdot 12H_2O$

1–2–a–6:	ice	a–2:	ice + 12h	d–e:	bi + ep
2–3–c–b–a:	12h	a–6:	ice + 12h*	e–f:	ep + ks
3–4–d–c:	8h	a–b:	12h + 12h*	e–g:	bi + ks
4–5–g–e–d:	bi	b–7:	ep + 12h*	f–8:	ep + hx
6–7–b–a:	12h*	b–c:	ep + 12h	f–9:	hx + ks
7–8–f–e–d–c–b:	ep	c–3:	8h + 12h		
8–9–f:	hx	c–d:	ep + 8h		
9–10–g–e–f:	ks	d–4:	bi + 8h		

	°C	{$MgCl_2$}	{$MgSO_4$}
1: ice + water	0	0	0
2: ice + 12h	–33.6	51.4	0
3: 8h + 12h	–16.8	87.5	0
4: bi + 8h	–3.4	99.0	0
6: ice + 12h*	–3.9	0	32.8
7: ep + 12h*	1.8	0	40.7
8: ep + hx	48.2	0	73.4
9: hx + ks	67.5	0	87.8
a: ice + 12h + 12h*	–35.0	51.0	1.0
b: ep + 12h + 12h*	–21.0	69.0	1.5
c: ep + 8h + 12h	–17.4	85.0	1.8
d: bi + ep + 8h	–3.8	97.2	2.5
e: bi + ep + ks	13.0	99.8	6.0
f: ep + hx + ks	18.0	83.5	7.5

SALSYS
Input: temperature. Output: {$MgCl_2$}, {$MgSO_4$}

		°C	
d–e: bi + ep	–3.8	–	13.0
e–g: bi + ks	13.0	–	100.0
f–8: ep + hx	18.0	–	48.2
f–9: hx + ks	18.0	–	67.5

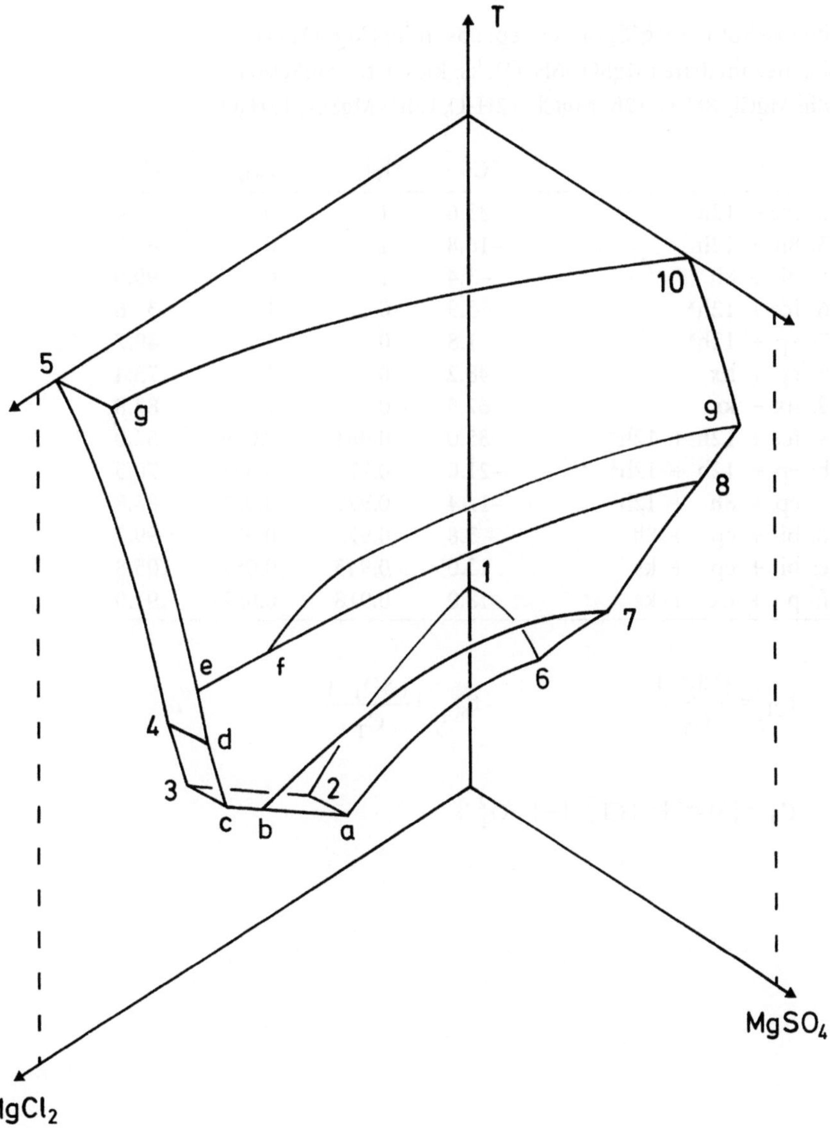
T
5
10
g
9
8
1
7
e
f
6
4
d
3
2
c
b
a
MgSO₄
MgCl₂

III.3.9
$MgCl_2$-$MgSO_4$-H_2O

bi: bischofite ($MgCl_2 \cdot 6H_2O$), ep: epsomite ($MgSO_4 \cdot 7H_2O$)
hx: hexahydrite ($MgSO_4 \cdot 6H_2O$), ks: kieserite ($MgSO_4 \cdot H_2O$)
8h: $MgCl_2 \cdot 8H_2O$, 12h: $MgCl_2 \cdot 12H_2O$, 12h*: $MgSO_4 \cdot 12H_2O$

	°C	f_{Cl_2}	f_{SO_4}	C_T
2: ice + 12h	−33.6	1	0	51.4
3: 8h + 12h	−16.8	1	0	87.5
4: bi + 8h	−3.4	1	0	99.0
6: ice + 12h*	−3.9	0	1	32.8
7: ep + 12h*	1.8	0	1	40.7
8: ep + hx	48.2	0	1	73.4
9: hx + ks	67.5	0	1	87.8
a: ice + 12h + 12h*	−35.0	0.981	0.019	52.0
b: ep + 12h + 12h*	−21.0	0.979	0.021	70.5
c: ep + 8h + 12h	−17.4	0.979	0.021	86.8
d: bi + ep + 8h	−3.8	0.975	0.025	99.7
e: bi + ep + ks	13.0	0.943	0.057	105.8
f: p + hx + ks	18.0	0.918	0.082	91.0

$$f_{Cl_2} = \frac{\{Cl_2^{2-}\}}{C_T} \qquad f_{SO_4} = \frac{\{SO_4^{2-}\}}{C_T}$$

$$C_T = \{Mg^{2+}\} = \{Cl_2^{2-}\} + \{SO_4^{2-}\}$$

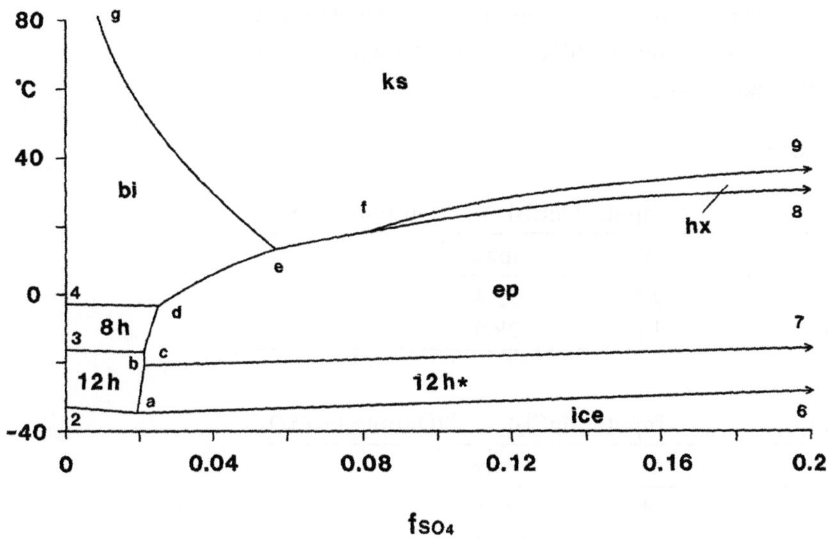
80
°C
40
0
-40
0
0.04
0.08
0.12
0.16
0.2
fSO4
g
ks
bi
9
f
8
hx
e
ep
4
d
8h
3
7
c
b
12h
12h*
a
ice
6
2

III.3.9
$MgCl_2$-$MgSO_4$-H_2O

bi: bischofite ($MgCl_2 \cdot 6H_2O$), ep: epsomite ($MgSO_4 \cdot 7H_2O$)
hx: hexahydrite ($MgSO_4 \cdot 6H_2O$), ks: kieserite ($MgSO_4 \cdot H_2O$)
12h*: $MgSO_4 \cdot 12H_2O$

SALSYS

ep	Input: {$MgCl_2$}			Output: {$MgSO_4$}
°C	min.		max.	
0	9.2	–	71.8	
25	4.1	–	39.0	

ks	Input: {$MgCl_2$}			Output: {$MgSO_4$}
°C	min.		max.	
75	14.4	–	103.1	
100	13.7	–	103.2	

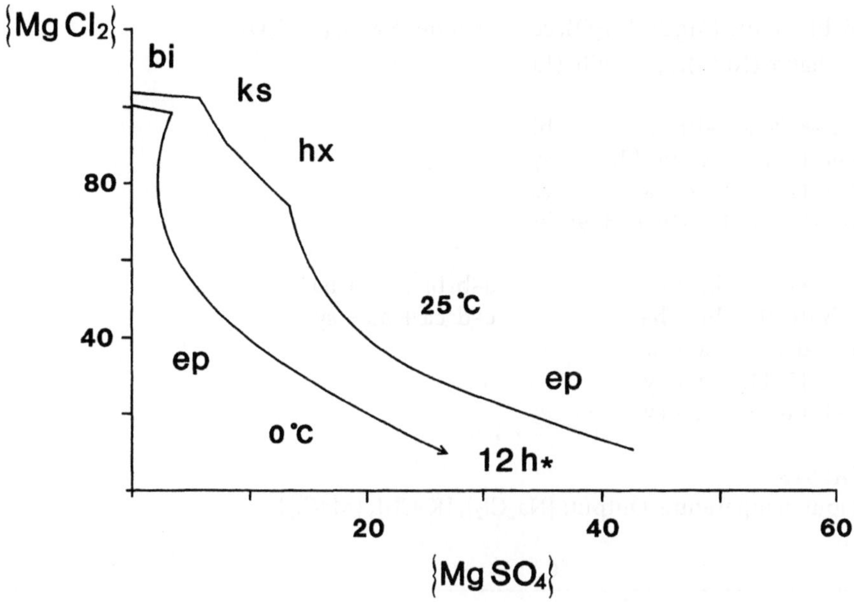
{Mg Cl2}
bi
ks
hx
80
40
25 °C
ep
ep
0 °C
12 h*
20
40
60
{Mg SO4}

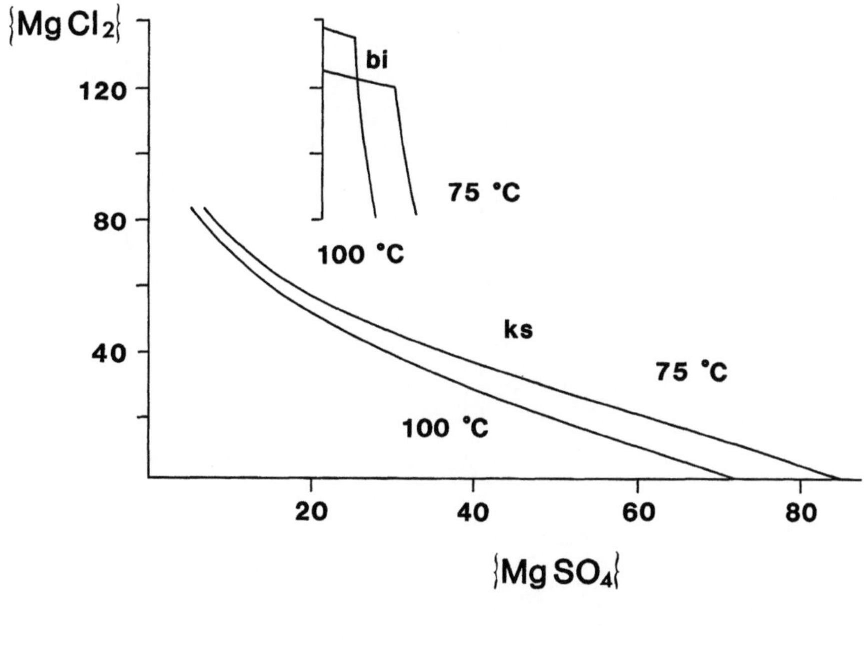
{Mg Cl2}
120
80
40
bi
75 °C
100 °C
ks
75 °C
100 °C
20
40
60
80
{Mg SO4}

III.4.1
Na_2Cl_2-K_2Cl_2-$MgCl_2$-H_2O

bi: bischofite ($MgCl_2 \cdot 6H_2O$), ca: carnallite ($KMgCl_3 \cdot 6H_2O$)
ha: halite (NaCl), sy: sylvite (KCl)

1–2–8–7–a–b–10–9:	bi
3–4–12–11–c–d–14–13:	sy
7–8–12–11–c–d–b–a:	ca
a–c–13–14–d–b–10–6–5–9:	ha

a–b–8–7:	bi + ca
a–b–10–9:	bi + ha
a–b–d–c:	ca + ha
c–d–12–11:	ca + sy
c–d–14–13:	ha + sy

a–b: bi + ca + ha
c–d: ca + ha + sy

SALSYS
Input: temperature. Output: {Na_2Cl_2}, {K_2Cl_2}, {$MgCl_2$}

	°C		
a–b: bi + ca + ha	0	–	100
c–d: ca + ha + sy	−10	–	110

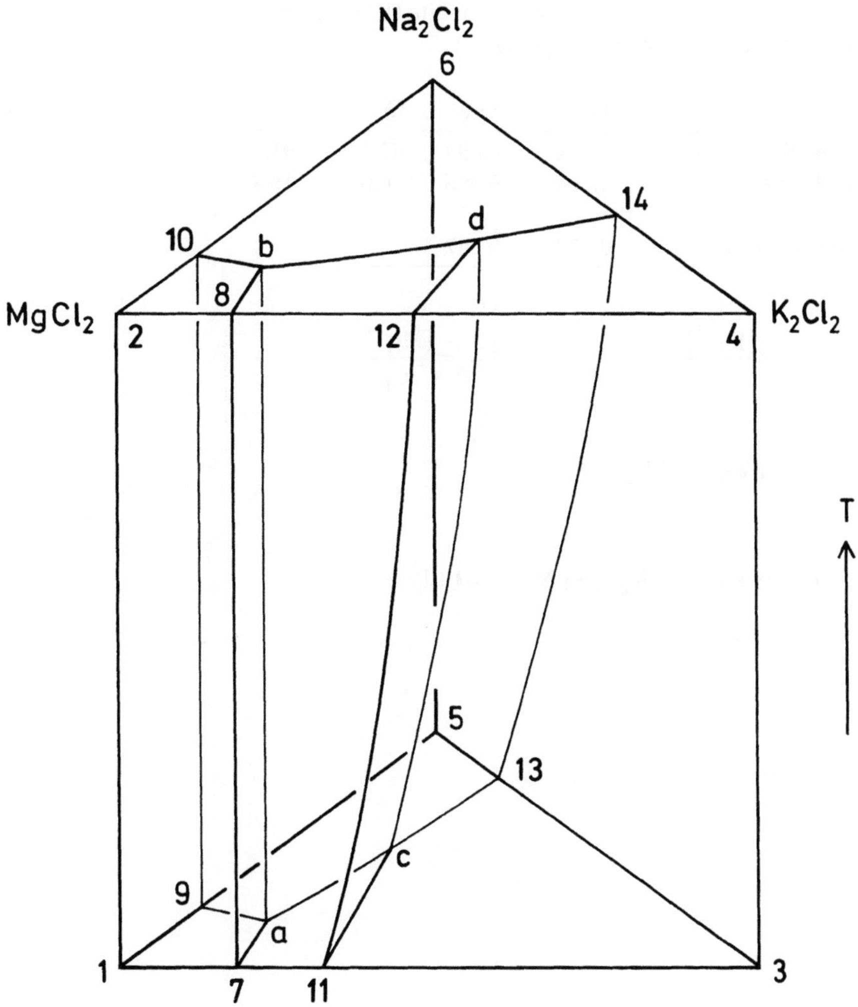
Na$_2$Cl$_2$
6
14
d
10
b
8
MgCl$_2$
2
12
4
K$_2$Cl$_2$
T
5
13
c
9
a
1
7
11
3

III.4.1
Na_2Cl_2-K_2Cl_2-$MgCl_2$-H_2O

bi: bischofite ($MgCl_2 \cdot 6H_2O$), ca: carnallite ($KMgCl_3 \cdot 6H_2O$)
ha: halite (NaCl), sy: sylvite (KCl)

60°C	F_{K_2}	F_{Mg}	C_P	C_T
a–b: bi + ca + ha	0.101	0.899	87.4	91.7
c–d: ca + ha + sy	0.004	0.996	116.0	116.9
salt point: ca	0.333	0.667		

$$F_{K_2} = \frac{\{K_2^{2+}\}}{C_P} \qquad F_{Mg} = \frac{\{Mg^{2+}\}}{C_P}$$

$$C_P = \{Mg^{2+}\} + \{K_2^{2+}\}$$

$$C_T = \{Mg^{2+}\} + \{K_2^{2+}\} + \{Na_2^{2+}\} = \{Cl_2^{2-}\}$$

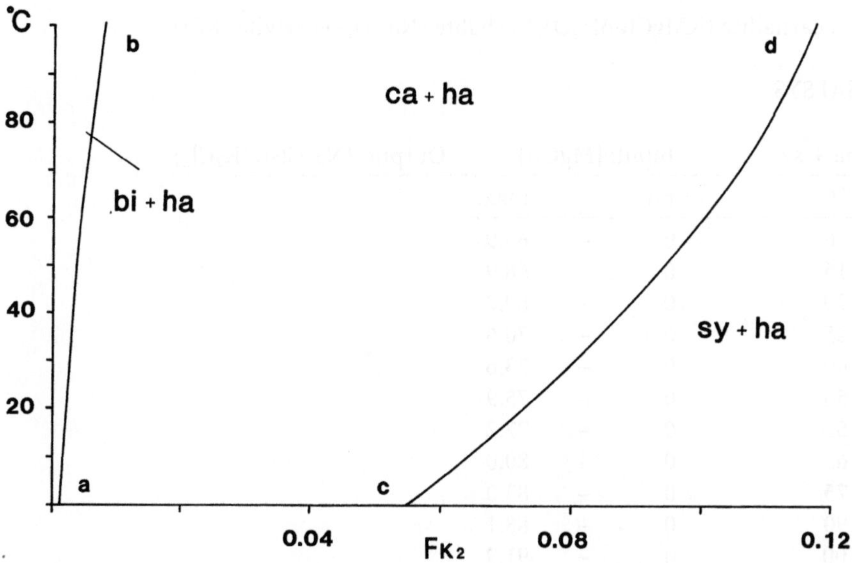
°C
80
60
40
20
0.04
0.08
0.12
F_{K_2}
a
b
c
d
ca + ha
bi + ha
sy + ha

III.4.1
Na_2Cl_2-K_2Cl_2-$MgCl_2$-H_2O

ca: carnallite ($KMgCl_3{\cdot}6H_2O$), ha: halite (NaCl), sy: sylvite (KCl)

SALSYS

ha + sy	Input: $\{MgCl_2\}$			Output: $\{Na_2Cl_2\}$, $\{K_2Cl_2\}$
°C	min.		max.	
0	0	–	66.9	
15	0	–	68.9	
20	0	–	69.7	
25	0	–	70.6	
40	0	–	73.6	
50	0	–	75.9	
55	0	–	77.2	
65	0	–	80.0	
75	0	–	83.0	
90	0	–	88.1	
100	0	–	91.9	
105	0	–	93.9	

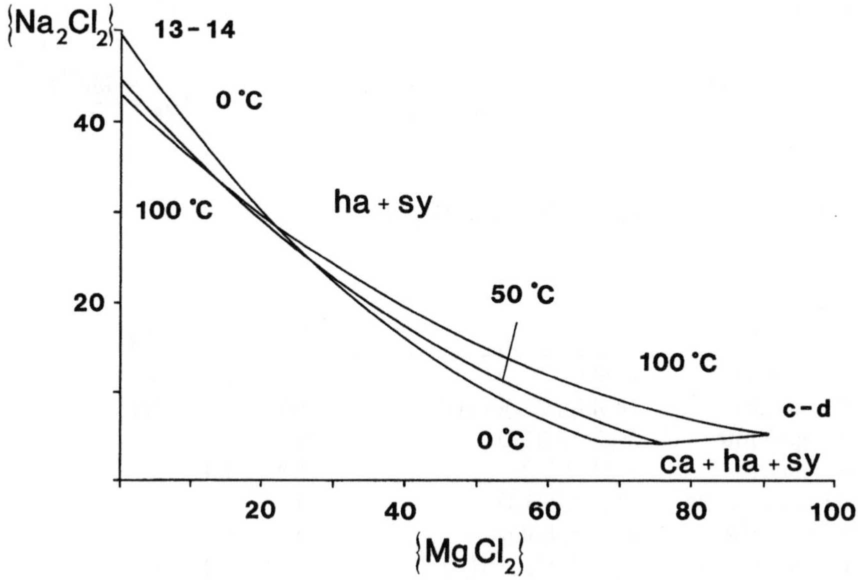

{Na2Cl2}
13 - 14
0 °C
40
100 °C
ha + sy
20
50 °C
100 °C
c - d
0 °C
ca + ha + sy
20
40
60
80
100
{Mg Cl2}

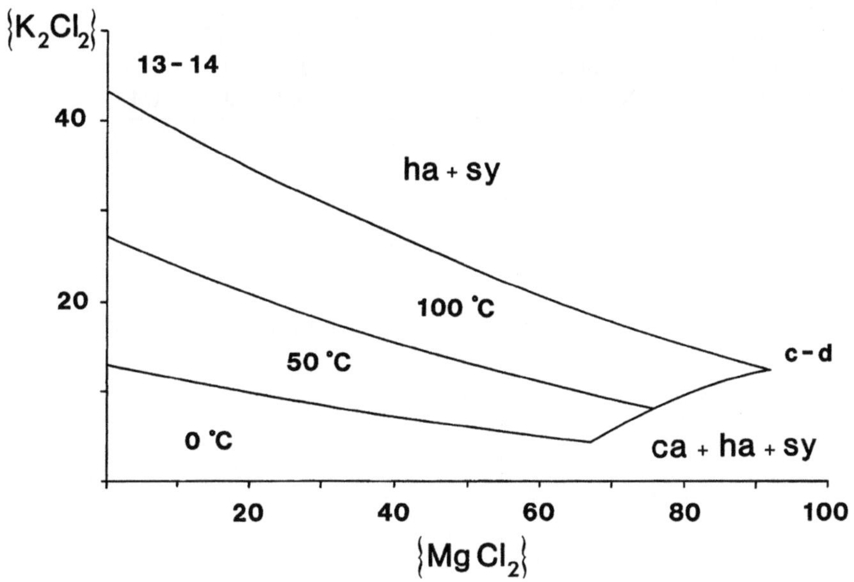

{K2Cl2}
13 - 14
40
ha + sy
20
100 °C
50 °C
c - d
0 °C
ca + ha + sy
20
40
60
80
100
{Mg Cl2}

III.4.2
Na_2SO_4-K_2SO_4-$MgSO_4$-H_2O

Solutions

$$f_{Na_2} = \frac{\{Na_2^{2+}\}}{C_T} \qquad f_{K_2} = \frac{\{K_2^{2+}\}}{C_T} \qquad f_{Mg} = \frac{\{Mg^{2+}\}}{C_T}$$

$$C_T = \{Na_2^{2+}\} + \{K_2^{2+}\} + \{Mg^{2+}\} = \{SO_4^{2-}\}$$

Salt points

		f_{Na_2}	f_{K_2}	f_{Mg}
ac: arcanite	K_2SO_4	0	1	0
bl: bloedite	$Na_2Mg(SO_4)_2 \cdot 4H_2O$	1/2	0	1/2
ep: epsomite	$MgSO_4 \cdot 7H_2O$	0	0	1
gs: glaserite	$K_3Na(SO_4)_2$	1/4	3/4	0
hx: hexahydrite	$MgSO_4 \cdot 6H_2O$	0	0	1
ks: kieserite	$MgSO_4 \cdot H_2O$	0	0	1
le: leonite	$K_2Mg(SO_4)_2 \cdot 4H_2O$	0	1/2	1/2
lg: langbeinite	$K_2Mg_2(SO_4)_3$	0	1/3	2/3
lw: loeweite	$Na_{12}Mg_7(SO_4)_{13} \cdot 15H_2O$	6/13	0	7/13
mi: mirabilite	$Na_2SO_4 \cdot 10H_2O$	1	0	0
pc: picromerite	$K_2Mg(SO_4)_2 \cdot 6H_2O$	0	1/2	1/2
th: thenardite	Na_2SO_4	1	0	0
vh: vanthoffite	$Na_6Mg(SO_4)_4$	3/4	0	1/4

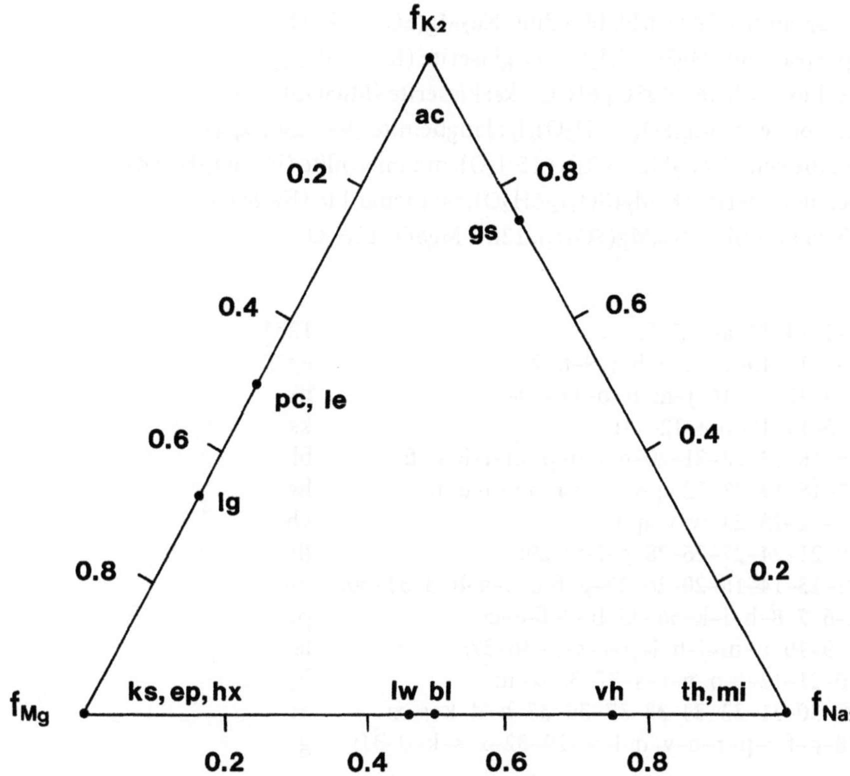

f_K2
ac
0.2
0.8
gs
0.4
0.6
pc, le
0.6
0.4
lg
0.8
0.2
f_Mg
ks, ep, hx
lw
bl
vh
th, mi
f_Na2
0.2
0.4
0.6
0.8

III.4.2
Na_2SO_4-K_2SO_4-$MgSO_4$-H_2O

ac: arcanite (K_2SO_4), bl: bloedite ($Na_2Mg(SO_4)_2 \cdot 4H_2O$)
ep: epsomite ($MgSO_4 \cdot 7H_2O$), gs: glaserite ($K_3Na(SO_4)_2$)
hx: hexahydrite ($MgSO_4 \cdot 6H_2O$), ks: kieserite ($MgSO_4 \cdot H_2O$)
le: leonite ($K_2Mg(SO_4)_2 \cdot 4H_2O$), lg: langbeinite ($K_2Mg_2(SO_4)_3$)
lw: loeweite ($Na_{12}Mg_7(SO_4)_{13} \cdot 15H_2O$), mi: mirabilite ($Na_2SO_4 \cdot 10H_2O$)
pc: picromerite ($K_2Mg(SO_4)_2 \cdot 6H_2O$), th: thenardite ($Na_2SO_4$)
vh: vanthoffite ($Na_6Mg(SO_4)_4$), 12h*: $MgSO_4 \cdot 12H_2O$

1–2–14–13–a–c–7–6:	12h*
2–3–16–15–14–c–e–h–j–9–8–7:	ep
3–4–18–17–16–j–m–n–o–11–10–9:	hx
4–5–19–18–o–t–12–11:	ks
15–16–17–22–21–20–g–l–q–p–m–j–h–e–f:	bl
17–18–19–23–22–q–v–u–r–p–m–n–o–t:	lw
21–22–23–24–w–v–q–l:	vh
20–21–24–27–26–28–g–l–w–29:	th
25–13–14–15–20–26–28–g–f–e–c–a–b–d–31–30:	mi
a–6–7–8–h–i–k–36–35–b–d–f–e–c:	pc
8–9–10–n–m–j–h–i–p–r–s–k–36–37:	le
10–11–12–t–o–n–r–s–37–38–x–u:	lg
33–30–31–32–34–38–37–36–35–b–d–k–s–x:	ac
28–g–f–i–p–r–u–v–q–l–w–29–32–x–s–k–d–31:	gs

Point 14: above point 13

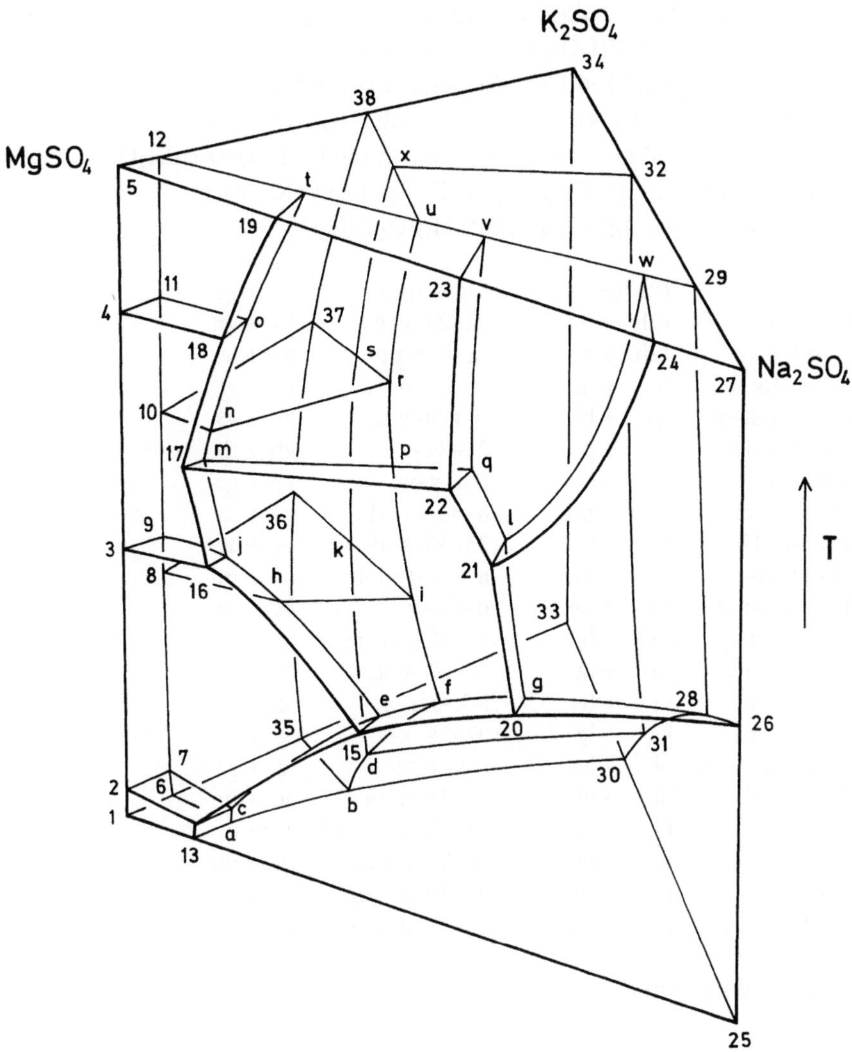
K2SO4
MgSO4
Na2SO4
T
1
2
3
4
5
6
7
8
9
10
11
12
13
15
16
17
18
19
20
21
22
23
24
25
26
27
28
29
30
31
32
33
34
35
36
37
38
a
b
c
d
e
f
g
h
i
j
k
l
m
n
o
p
q
r
s
t
u
v
w
x

III.4.2
Na_2SO_4-K_2SO_4-$MgSO_4$-H_2O

ac: arcanite (K_2SO_4), bl: bloedite ($Na_2Mg(SO_4)_2{\cdot}4H_2O$)
ep: epsomite ($MgSO_4{\cdot}7H_2O$), gs: glaserite ($K_3Na(SO_4)_2$)
hx: hexahydrite ($MgSO_4{\cdot}6H_2O$), ks: kieserite ($MgSO_4{\cdot}H_2O$)
le: leonite ($K_2Mg(SO_4)_2{\cdot}4H_2O$), lg: langbeinite ($K_2Mg_2(SO_4)_3$)
lw: loeweite ($Na_{12}Mg_7(SO_4)_{13}{\cdot}15H_2O$), mi: mirabilite ($Na_2SO_4{\cdot}10H_2O$)
pc: picromerite ($K_2Mg(SO_4)_2{\cdot}6H_2O$), th: thenardite ($Na_2SO_4$)
vh: vanthoffite ($Na_6Mg(SO_4)_4$), 12h*: $MgSO_4{\cdot}12H_2O$

2–7–c–14:	12h* + ep	f–g–l–q–p–i:	bl + gs
6–7–c–a:	12h* + pc	22–23–v–q:	lw + vh
13–14–c–a:	12h* + mi	m–n–r–p:	lw + le
3–9–j–16:	ep + hx	n–o–t–u–r:	lw + lg
15–16–j–h–e:	ep + bl	p–r–u–v–q:	lw + gs
7–8–h–e–c:	ep + pc	21–24–w–l:	vh + th
8–9–j–h:	ep + le	l–q–v–w:	vh + gs
14–15–e–c:	ep + mi	8–36–k–i–h:	pc + le
4–11–o–18:	hx + ks	35–36–k–d–b:	pc + ac
16–17–m–j:	hx + bl	a–c–e–f–d–b:	pc + mi
17–18–o–n–m:	hx + lw	d–f–i–k:	pc + gs
9–10–n–m–j:	hx + le	10–37–s–r–n:	le + lg
10–11–o–n:	hx + lg	36–37–s–k:	le + ac
18–19–t–o:	ks + lw	i–k–s–r–p:	le + gs
11–12–t–o:	ks + lg	37–38–x–s:	lg + ac
17–22–q–p–m:	bl + lw	r–s–x–u:	lg + gs
21–22–q–l:	bl + vh	30–31–d–b:	ac + mi
e–f–i–h:	bl + pc	31–32–x–s–k–d:	ac + gs
h–i–p–m–j:	bl + le	20–26–28–g:	th + mi
20–21–l–g:	bl + th	28–29–w–l–g:	th + gs
15–20–g–f–e:	bl + mi	28–31–d–f–g:	mi + gs

Point 14: above point 13

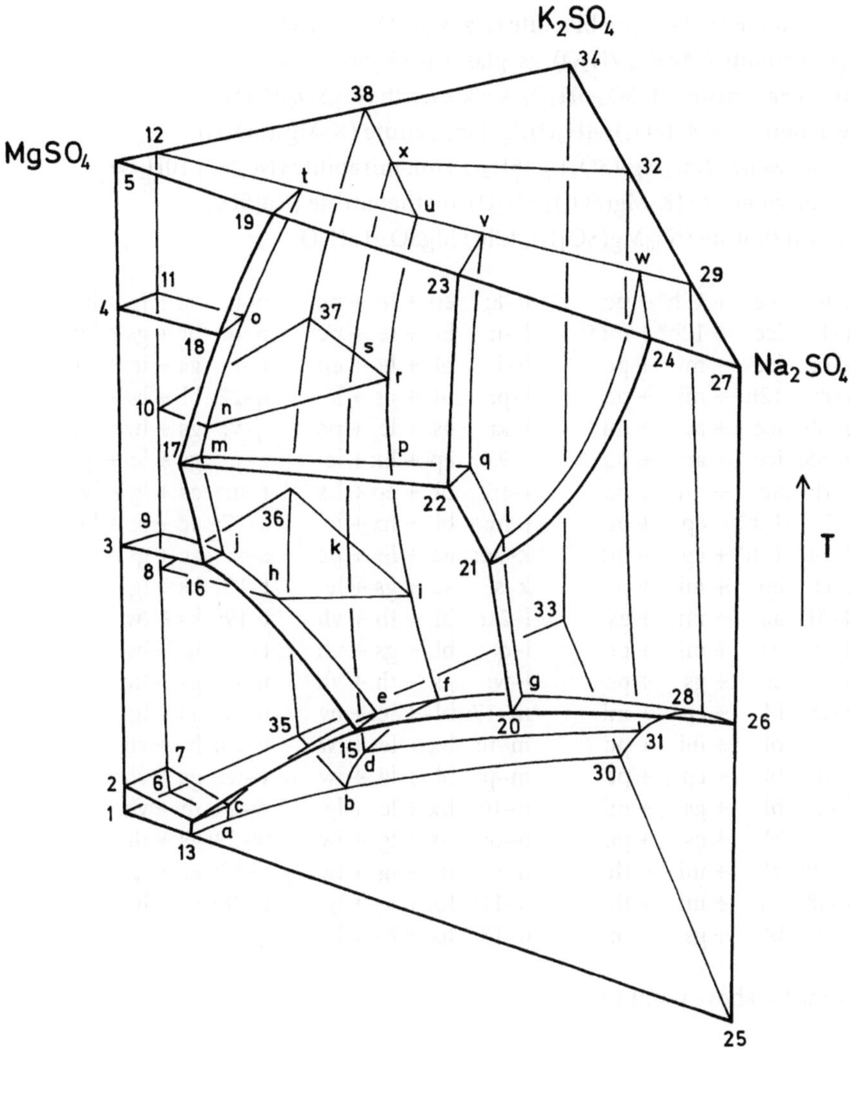

K_2SO_4
$MgSO_4$
Na_2SO_4
T

III.4.2
Na_2SO_4-K_2SO_4-$MgSO_4$-H_2O

ac: arcanite (K_2SO_4), bl: bloedite ($Na_2Mg(SO_4)_2 \cdot 4H_2O$)
ep: epsomite ($MgSO_4 \cdot 7H_2O$), gs: glaserite ($K_3Na(SO_4)_2$)
hx: hexahydrite ($MgSO_4 \cdot 6H_2O$), ks: kieserite ($MgSO_4 \cdot H_2O$)
le: leonite ($K_2Mg(SO_4)_2 \cdot 4H_2O$), lg: langbeinite ($K_2Mg_2(SO_4)_3$)
lw: loeweite ($Na_{12}Mg_7(SO_4)_{13} \cdot 15H_2O$), mi: mirabilite ($Na_2SO_4 \cdot 10H_2O$)
pc: picromerite ($K_2Mg(SO_4)_2 \cdot 6H_2O$), th: thenardite ($Na_2SO_4$)
vh: vanthoffite ($Na_6Mg(SO_4)_4$), 12h*: $MgSO_4 \cdot 12H_2O$

a–6: ice + 12h* + pc
a–13: ice + 12h* + mi
a–b: ice + mi + pc
a–c: 12h* + mi + pc
b–30: ice + ac + mi
b–35: ice + ac + pc
b–d: ac + mi + pc
c–7: 12h* + ep + pc
c–14: 12h* + ep + mi
c–e: ep + mi + pc
d–31: ac + mi + gs
d–f: gs + mi + pc
d–k: ac + gs + pc
e–15: bl + ep + mi
e–f: bl + mi + pc
e–h: bl + ep + pc
f–g: bl + gs + mi
f–i: bl + gs + pc
g–20: bl + mi + th
g–28: gs + mi + th
g–l: bl + gs + th
h–8: ep + le + pc
h–i: bl + le + pc
h–j: bl + le + ep
i–p: bl + gs + le
i–k: gs + le + pc
j–9: ep + hx + le
j–16: bl + ep + hx
j–m: bl + hx + le
k–36: ac + le + pc
k–s: ac + gs + le
l–21: bl + th + vh
l–q: bl + gs + vh
l–w: gs + th + vh
m–17: bl + hx + lw
m–n: hx + le + lw
m–p: bl + le + lw
n–10: hx + le + lg
n–o: hx + lg + lw
n–r: le + lg + lw
o–11: hx + ks + lg
o–18: hx + ks + lw
o–t: ks + lg + lw
p–q: bl + gs + lw
p–r: gs + le + lw
q–22: bl + lw + vh
q–v: gs + lw + vh
r–s: gs + le + lg
r–u: gs + lg + lw
s–37: ac + le + lg
s–x: ac + gs + lg
t–12: ks + lg
t–19: ks + lw
t–u: lg + lw
u–v: gs + lw
u–x: gs + lg
v–23: lw + vh
v–w: gs + vh
w–24: th + vh
w–29: gs + th
x–32: ac + gs
x–38: ac + lg

Point 14: above point 13

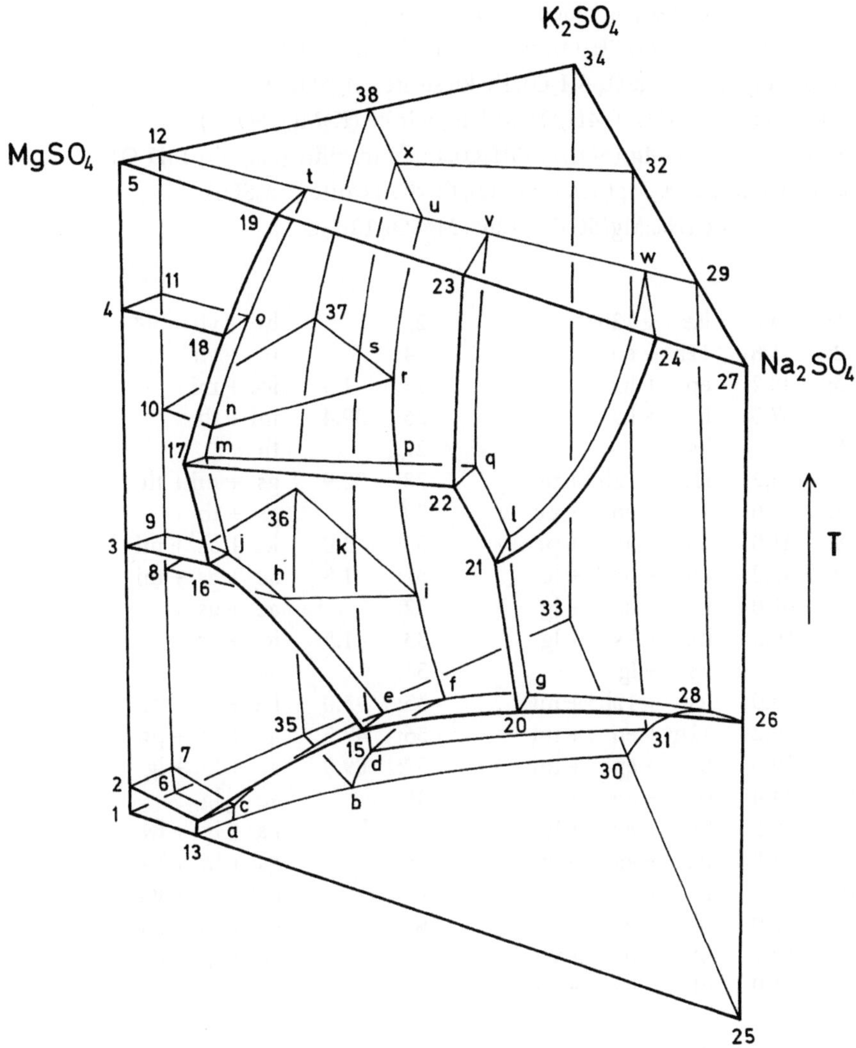
K2SO4
MgSO4
Na2SO4
T

III.4.2
Na_2SO_4-K_2SO_4-$MgSO_4$-H_2O

ac: arcanite (K_2SO_4), bl: bloedite ($Na_2Mg(SO_4)_2{\cdot}4H_2O$)
ep: epsomite ($MgSO_4{\cdot}7H_2O$), gs: glaserite ($K_3Na(SO_4)_2$)
hx: hexahydrite ($MgSO_4{\cdot}6H_2O$), ks: kieserite ($MgSO_4{\cdot}H_2O$)
le: leonite ($K_2Mg(SO_4)_2{\cdot}4H_2O$), lg: langbeinite ($K_2Mg_2(SO_4)_3$)
lw: loeweite ($Na_{12}Mg_7(SO_4)_{13}{\cdot}15H_2O$), mi: mirabilite ($Na_2SO_4{\cdot}10H_2O$)
pc: picromerite ($K_2Mg(SO_4)_2{\cdot}6H_2O$), th: thenardite ($Na_2SO_4$)
vh: vanthoffite ($Na_6Mg(SO_4)_4$), 12h*: $MgSO_4{\cdot}12H_2O$

	°C			°C	
1:	−3.9	ice + 12h*	23:		lw + vh
2:	1.8	12h* + ep	24:		th + vh
3:	48.2	ep + hx	25:	−1.3	ice + mi
4:	67.5	hx + ks	26:	32.4	mi + th
5:		ks	27:		th
6:	−5.2	ice + 12h* + pc	28:	30.9	gs + mi + th
7:	1.0	12h* + ep + pc	29:		gs + th
8:	41.0	ep + le + pc	30:	−3.0	ice + ac + mi
9:	47.2	ep + hx + le	31:	1.8	ac + gs + mi
10:	61.0	hx + le + lg	32:		ac + gs
11:	66.5	hx + ks + lg	33:	−1.5	ice + ac
12:		ks + lg	34:		ac
13:	−5.0	ice + 12h* + mi	35:	−3.0	ice + ac + pc
14:	0.2	12h* + ep + mi	36:	47.5	ac + le + pc
15:	20.6	bl + ep + mi	37:	89.0	ac + le + lg
16:	44.4	bl + ep + hx	38:		ac + lg
17:	59.5	bl + hx + lw	t:		ks + lg + lw
18:	64.0	hx + ks + lw	u:		gs + lg + lw
19:		ks + lw	v:		gs + lw + vh
20:	27.0	bl + mi + th	w:		gs + th + vh
21:	60.0	bl + th + vh	x:		ac + gs + lg
22:	71.0	bl + lw + vh			

Point 14: above point 13

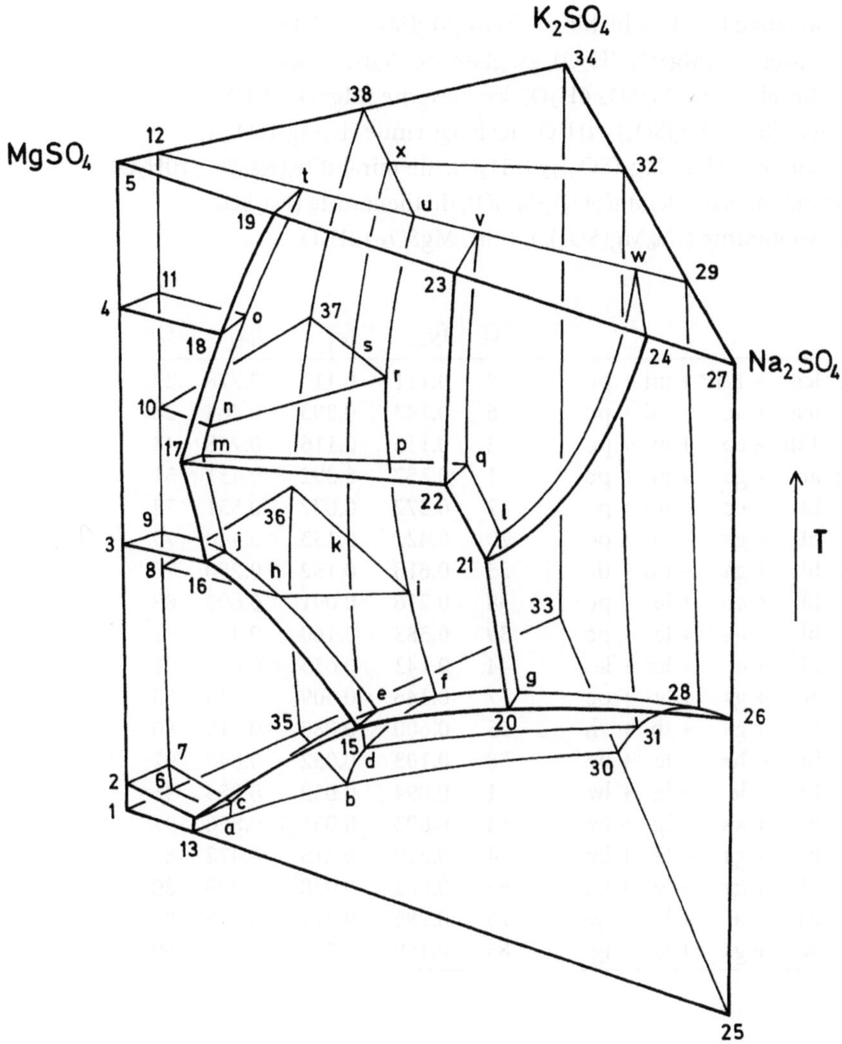

K₂SO₄
MgSO₄
Na₂SO₄
T
1
2
3
4
5
6
7
8
9
10
11
12
13
15
16
17
18
19
20
21
22
23
24
25
26
27
28
29
30
31
32
33
34
35
36
37
38
a
b
c
d
e
f
g
h
i
j
k
l
m
n
o
p
q
r
s
t
u
v
w
x

III.4.2
Na_2SO_4-K_2SO_4-$MgSO_4$-H_2O

ac: arcanite (K_2SO_4), bl: bloedite ($Na_2Mg(SO_4)_2{\cdot}4H_2O$)
ep: epsomite ($MgSO_4{\cdot}7H_2O$), gs: glaserite ($K_3Na(SO_4)_2$)
hx: hexahydrite ($MgSO_4{\cdot}6H_2O$), ks: kieserite ($MgSO_4{\cdot}H_2O$)
le: leonite ($K_2Mg(SO_4)_2{\cdot}4H_2O$), lg: langbeinite ($K_2Mg_2(SO_4)_3$)
lw: loeweite ($Na_{12}Mg_7(SO_4)_{13}{\cdot}15H_2O$), mi: mirabilite ($Na_2SO_4{\cdot}10H_2O$)
pc: picromerite ($K_2Mg(SO_4)_2{\cdot}6H_2O$), th: thenardite ($Na_2SO_4$)
vh: vanthoffite ($Na_6Mg(SO_4)_4$), 12h*: $MgSO_4{\cdot}12H_2O$

	°C	f_{Na_2}	f_{K_2}	f_{Mg}	C_T
a: ice + 12h* + mi + pc	−8	0.111	0.111	0.778	27
b: ice + ac + mi + pc	−6	0.143	0.393	0.464	28
c: 12h* + ep + mi + pc	−2	0.116	0.116	0.768	43
d: ac + gs + mi + pc	−1	0.157	0.392	0.451	51
e: bl + ep + mi + pc	17	0.372	0.077	0.551	78
f: bl + gs + mi + pc	20	0.427	0.133	0.440	75
g: bl + gs + mi + th	25	0.618	0.132	0.250	76
h: bl + ep + le + pc	38	0.216	0.091	0.693	88
i: bl + gs + le + pc	39	0.383	0.160	0.457	81
j: bl + ep + hx + le	44	0.143	0.055	0.802	91
k: ac + gs + le + pc	7	0.146	0.309	0.545	55
l: bl + gs + th + vh	55	0.600	0.088	0.312	80
m: bl + hx + le + lw	58	0.103	0.052	0.845	97
n: hx + le + lg + lw	1	0.094	0.052	0.854	96
o: hx + ks + lg + lw	64	0.093	0.051	0.856	97
p: bl + gs + le + lw	64	0.310	0.218	0.472	87
q: bl + gs + lw + vh	66	0.562	0.038	0.400	80
r: gs + le + lg + lw	75	0.286	0.214	0.500	84
s: ac + gs + le + lg	85	0.159	0.290	0.551	69

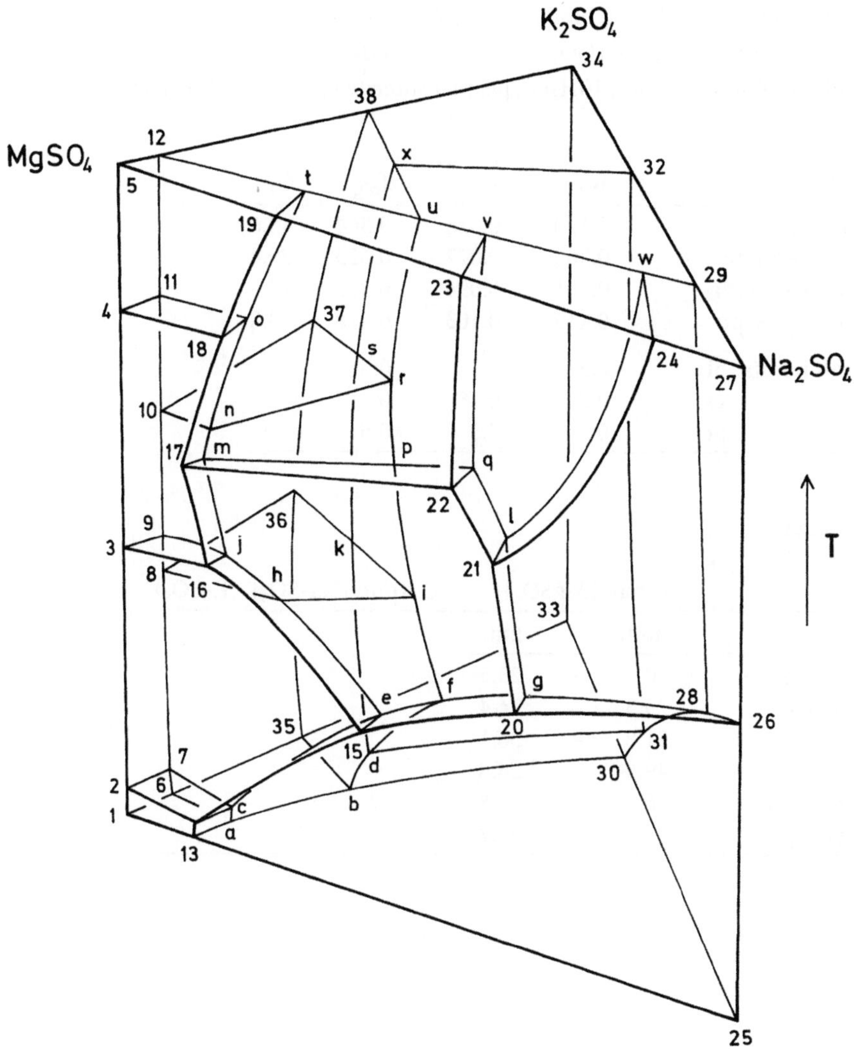

K2SO4
MgSO4
Na2SO4
T

III.4.2
Na_2SO_4-K_2SO_4-$MgSO_4$-H_2O

ac: arcanite (K_2SO_4), bl: bloedite ($Na_2Mg(SO_4)_2 \cdot 4H_2O$)
ep: epsomite ($MgSO_4 \cdot 7H_2O$), gs: glaserite ($K_3Na(SO_4)_2$)
mi: mirabilite ($Na_2SO_4 \cdot 10H_2O$), pc: picromerite ($K_2Mg(SO_4)_2 \cdot 6H_2O$)

25°C	f_{Na_2}	f_{K_2}	f_{Mg}	C_T
ac + gs + pc	0.170	0.326	0.504	46.6
bl + ep + pc	0.300	0.077	0.623	79.4
bl + gs + mi	0.501	0.094	0.405	78.6
bl + gs + pc	0.419	0.103	0.478	78.8
salt points: bl	0.5	0	0.5	
gs	0.25	0.75	0	
pc	0	0.5	0.5	

SALSYS

	Input: {$MgSO_4$}			Output: {Na_2SO_4}, {K_2SO_4}
	min.		max.	
ac + gs	0	–	23.5	
ac + pc	23.5	–	24.4	
bl + mi	31.8	–	36.1	
ep + pc	49.5	–	56.8	
gs + mi	0	–	31.8	
gs + pc	23.5	–	37.7	

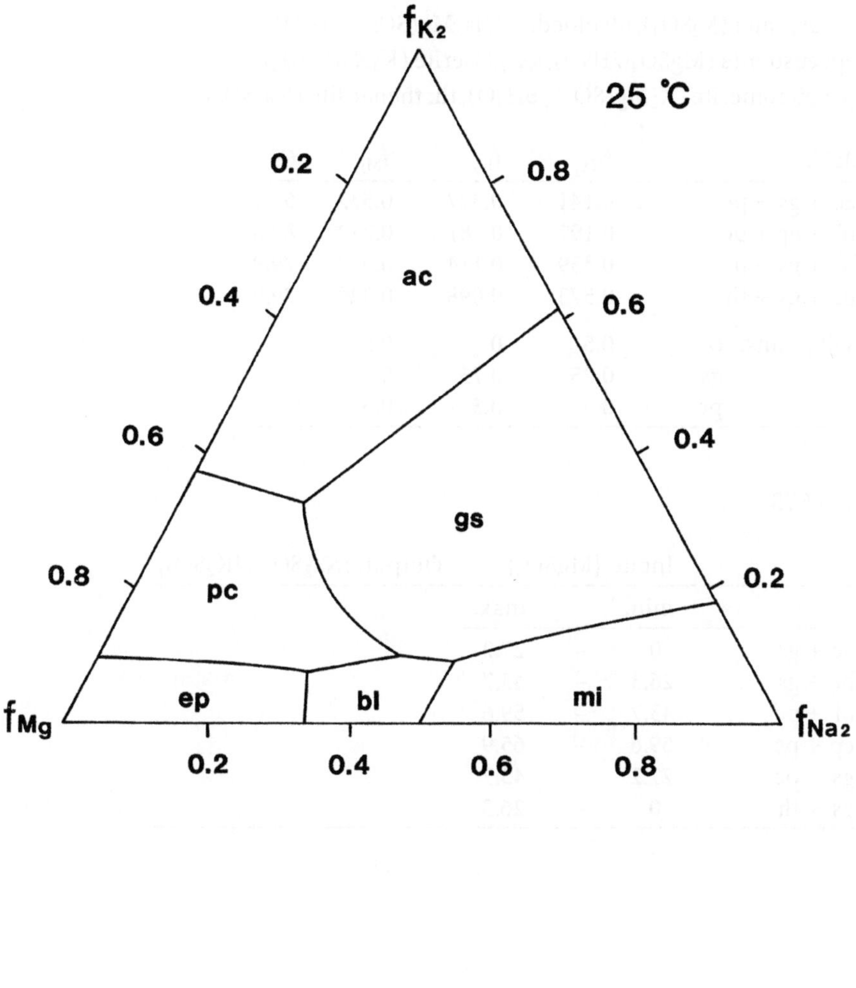

f K2
25 °C
0.2
0.8
0.4
0.6
0.6
0.4
0.8
0.2
ac
gs
pc
ep
bl
mi
f Mg
f Na2
0.2
0.4
0.6
0.8

III.4.2
Na_2SO_4-K_2SO_4-$MgSO_4$-H_2O

ac: arcanite (K_2SO_4), bl: bloedite ($Na_2Mg(SO_4)_2{\cdot}4H_2O$)
ep: epsomite ($MgSO_4{\cdot}7H_2O$), gs: glaserite ($K_3Na(SO_4)_2$)
pc: picromerite ($K_2Mg(SO_4)_2{\cdot}6H_2O$), th: thenardite ($Na_2SO_4$)

35°C		f_{Na_2}	f_{K_2}	f_{Mg}	C_T
ac + gs + pc		0.141	0.327	0.532	51.1
bl + ep + pc		0.197	0.081	0.722	82.5
bl + gs + pc		0.339	0.114	0.547	79.9
bl + gs + th		0.573	0.098	0.329	79.9
salt points:	bl	0.5	0	0.5	
	gs	0.25	0.75	0	
	pc	0	0.5	0.5	

SALSYS

	Input: $\{MgSO_4\}$			Output: $\{Na_2SO_4\}$, $\{K_2SO_4\}$
	min.		max.	
ac + gs	0	–	27.2	
bl + gs	26.3	–	43.7	
bl + pc	43.7	–	59.6	
ep + pc	59.6	–	65.9	
gs + pc	27.2	–	43.7	
gs + th	0	–	26.3	

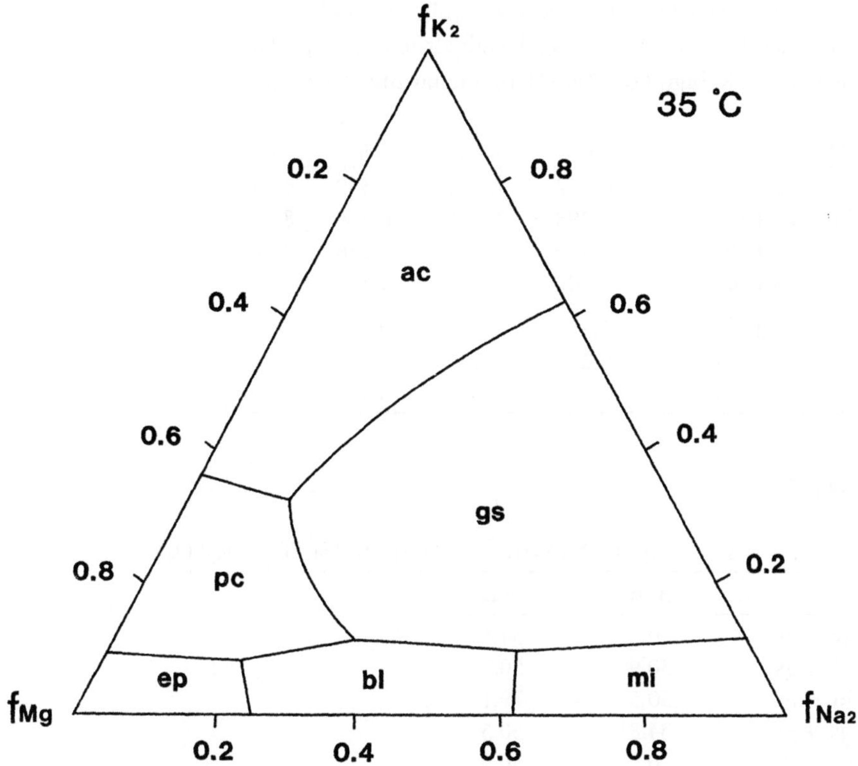

f_{K_2}
35 °C
0.2
0.4
0.6
0.8
0.8
0.6
0.4
0.2
ac
gs
pc
ep
bl
mi
f_{Mg}
f_{Na_2}
0.2
0.4
0.6
0.8

III.4.2
Na_2SO_4-K_2SO_4-$MgSO_4$-H_2O

ac: arcanite (K_2SO_4), bl: bloedite ($Na_2Mg(SO_4)_2{\cdot}4H_2O$)
gs: glaserite ($K_3Na(SO_4)_2$), hx: hexahydrite ($MgSO_4{\cdot}6H_2O$)
le: leonite ($K_2Mg(SO_4)_2{\cdot}4H_2O$), th: thenardite ($Na_2SO_4$)

55°C		f_{Na_2}	f_{K_2}	f_{Mg}	C_T
ac + gs + le		0.122	0.319	0.559	62.3
bl + gs + le		0.298	0.138	0.564	89.5
bl + gs + th		0.523	0.121	0.356	86.7
bl + hx + le		0.107	0.067	0.826	95.7
salt points:	bl	0.5	0	0.5	
	gs	0.25	0.75	0	
	le	0	0.5	0.5	

SALSYS

	Input: $\{MgSO_4\}$			Output: $\{Na_2SO_4\}$, $\{K_2SO_4\}$
	min.		max.	
ac + gs	0	–	34.8	
bl + gs	30.9	–	50.5	
bl + le	50.5	–	79.1	
gs + le	34.8	–	50.5	
gs + th	0	–	30.9	

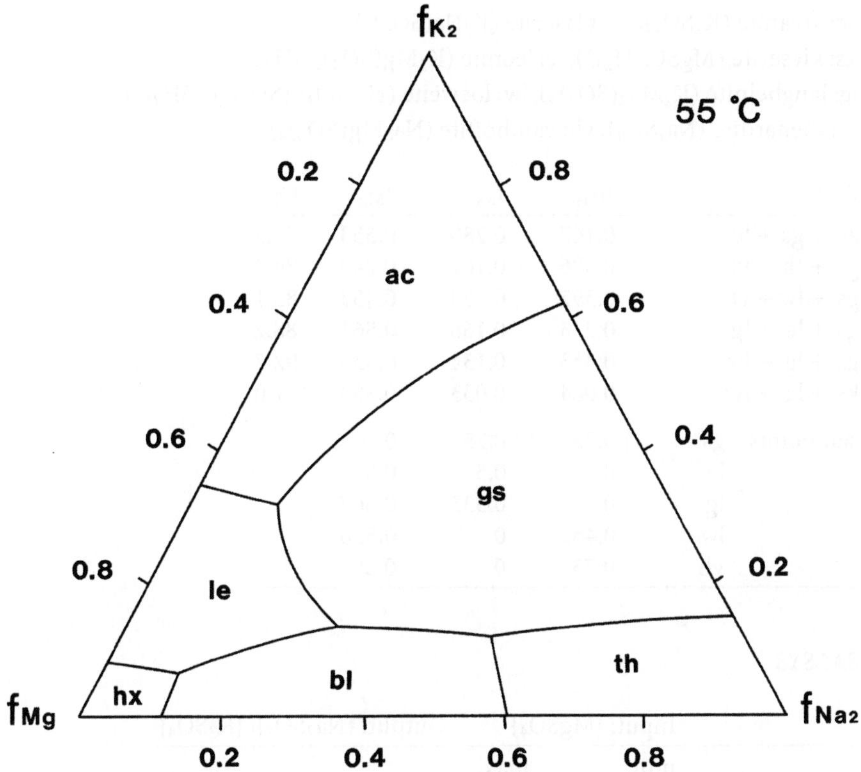

f_{K_2}
55 °C
0.2
0.8
ac
0.4
0.6
0.6
0.4
gs
0.8
le
0.2
th
bl
hx
f_{Mg}
f_{Na_2}
0.2
0.4
0.6
0.8

III.4.2
Na_2SO_4-K_2SO_4-$MgSO_4$-H_2O

ac: arcanite (K_2SO_4), gs: glaserite ($K_3Na(SO_4)_2$)
ks: kieserite ($MgSO_4 \cdot H_2O$), le: leonite ($K_2Mg(SO_4)_2 \cdot 4H_2O$)
lg: langbeinite ($K_2Mg_2(SO_4)_3$), lw: loeweite ($Na_{12}Mg_7(SO_4)_{13} \cdot 15H_2O$)
th: thenardite (Na_2SO_4), vh: vanthoffite ($Na_6Mg(SO_4)_4$)

75°C		f_{Na_2}	f_{K_2}	f_{Mg}	C_T
ac + gs + le		0.167	0.280	0.553	73.2
gs + th + vh		0.596	0.161	0.243	74.4
gs + lw + vh		0.397	0.150	0.453	83.1
gs + le + lg		0.273	0.166	0.561	88.2
gs + lg + lw		0.333	0.134	0.533	92.7
ks + lg + lw		0.084	0.033	0.883	111.0
salt points:	gs	0.25	0.75	0	
	le	0	0.5	0.5	
	lg	0	0.333	0.667	
	lw	0.462	0	0.538	
	vh	0.75	0	0.25	

SALSYS

	Input: $\{MgSO_4\}$			Output: $\{Na_2SO_4\}$, $\{K_2SO_4\}$
	min.		max.	
ac + gs	0	–	40.5	
gs + le	40.5	–	49.5	
gs + th	0	–	18.1	
lg + lw	49.4	–	98.0	

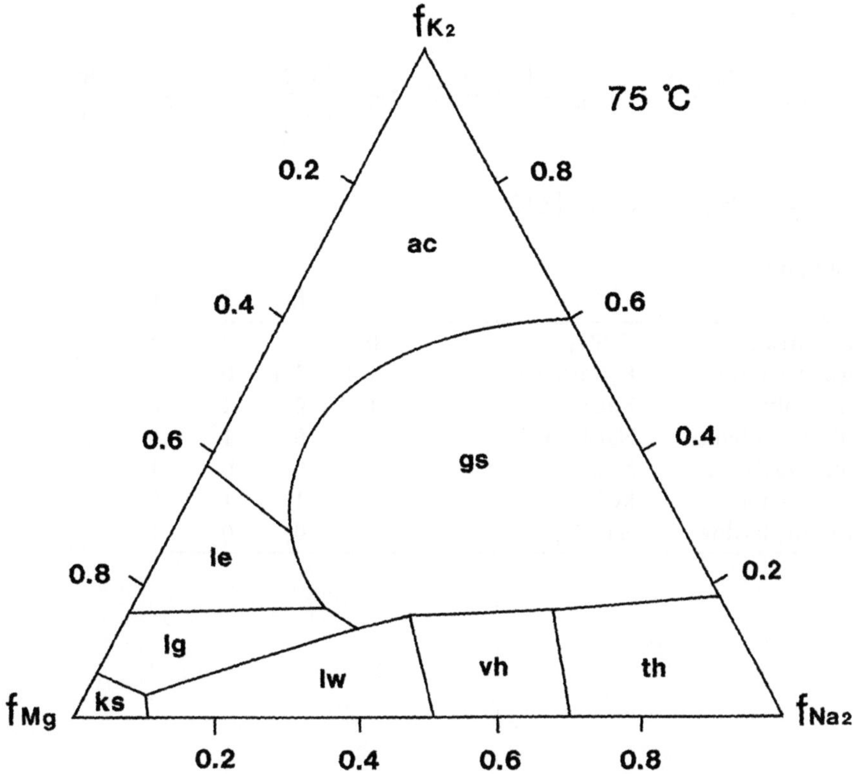

fK2
75 ℃
0.2
0.8
ac
0.4
0.6
gs
0.6
0.4
le
0.8
0.2
lg
vh
th
lw
ks
fMg
fNa2
0.2
0.4
0.6
0.8

III.4.3
Na_2Cl_2-K_2Cl_2-Na_2SO_4-K_2SO_4-H_2O

Solutions

$$f_{Na_2} = \frac{\{Na_2^{2+}\}}{C_T} \qquad f_{K_2} = \frac{\{K_2^{2+}\}}{C_T} \qquad f_{Cl_2} = \frac{\{Cl_2^{2-}\}}{C_T} \qquad f_{SO_4} = \frac{\{SO_4^{2-}\}}{C_T}$$

$$C_T = \{Na_2^{2+}\} + \{K_2^{2+}\} = \{Cl_2^{2-}\} + \{SO_4^{2-}\}$$

Salt points

		f_{Na_2}	f_{K_2}	f_{Cl_2}	f_{SO_4}
ac: arcanite	K_2SO_4	0	1	0	1
gs: glaserite	$K_3Na(SO_4)_2$	1/4	3/4	0	1
ha: halite	$NaCl$	1	0	1	0
hh: hydrohalite	$NaCl \cdot 2H_2O$	1	0	1	0
mi: mirabilite	$Na_2SO_4 \cdot 10H_2O$	1	0	0	1
sy: sylvite	KCl	0	1	1	0
th: thenardite	Na_2SO_4	1	0	0	1

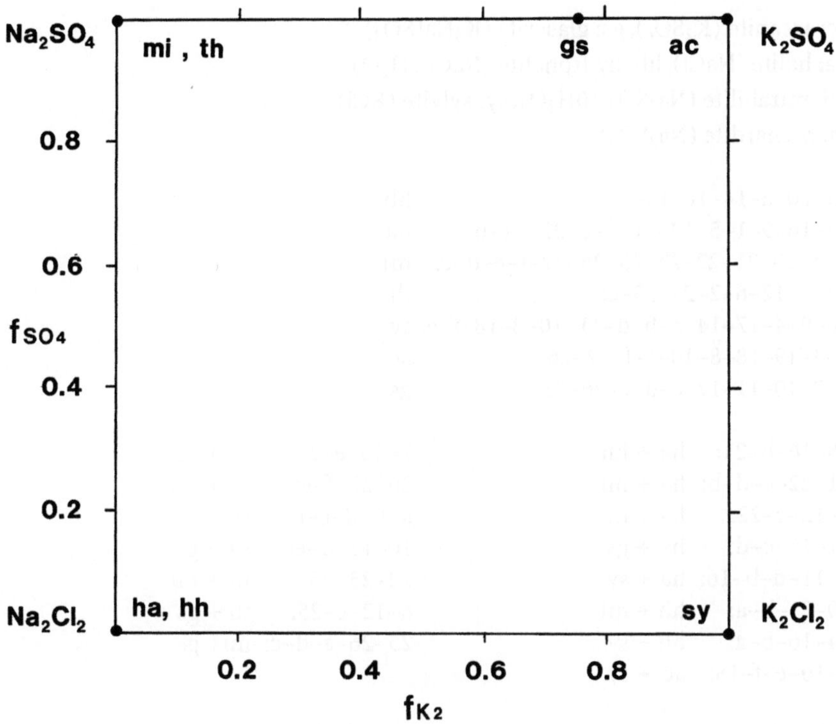

Na_2SO_4
mi , th
gs
ac
K_2SO_4
0.8
0.6
f_{SO_4}
0.4
0.2
Na_2Cl_2
ha, hh
sy
K_2Cl_2
0.2
0.4
0.6
0.8
f_{K_2}

III.4.3
Na_2Cl_2-K_2Cl_2-Na_2SO_4-K_2SO_4-H_2O

ac: arcanite (K_2SO_4), gs: glaserite ($K_3Na(SO_4)_2$)
ha: halite (NaCl), hh: hydrohalite ($NaCl{\cdot}2H_2O$)
mi: mirabilite ($Na_2SO_4{\cdot}10H_2O$), sy: sylvite (KCl)
th: thenardite (Na_2SO_4)

13–20–a–14–16–15–21–b:	hh
15–16–9–1–5–12–11–d–c–22–21–b:	ha
b–a–20–21–22–23–25–26–27–f–e–d–c:	mi
22–5–12–6–2–23–25–c:	th
16–9–4–17–14–a–b–d–11–10–8–18–f–e:	sy
7–3–19–18–8–10–e–f–27–26:	ac
6–7–10–11–12–c–d–e–26–25:	gs

15–16–b–21:	ha + hh	7–10–e–26:	ac + gs
21–22–c–d–b:	ha + mi	26–27–f–e:	ac + mi
5–12–c–22:	ha + th	a–b–d–e–f:	sy + mi
11–12–c–d:	ha + gs	10–11–d–e:	sy + gs
9–11–d–b–16:	ha + sy	22–23–25–c:	th + mi
20–21–b–a:	hh + mi	6–12–c–25:	th + gs
14–16–b–a:	hh + sy	25–26–e–d–c:	mi+ gs
8–10–e–f–18:	ac + sy		

a–14:	ice + hh + sy	c–25:	gs + mi + th
a–20:	ice + hh + mi	c–d:	gs + ha + mi
a–b:	hh + mi + sy	d–11:	gs + ha + sy
a–f:	ice + mi + sy	d–e:	gs + mi + sy
b–16:	ha + hh + sy	e–10:	ac + gs + sy
b–21:	ha + hh + mi	e–26:	ac + gs + mi
b–d:	ha + mi + sy	e–f:	ac + mi + sy
c–12:	ha + gs + th	f–18:	ice + ac + sy
c–22:	ha + mi + th	f–27:	ice + ac + mi

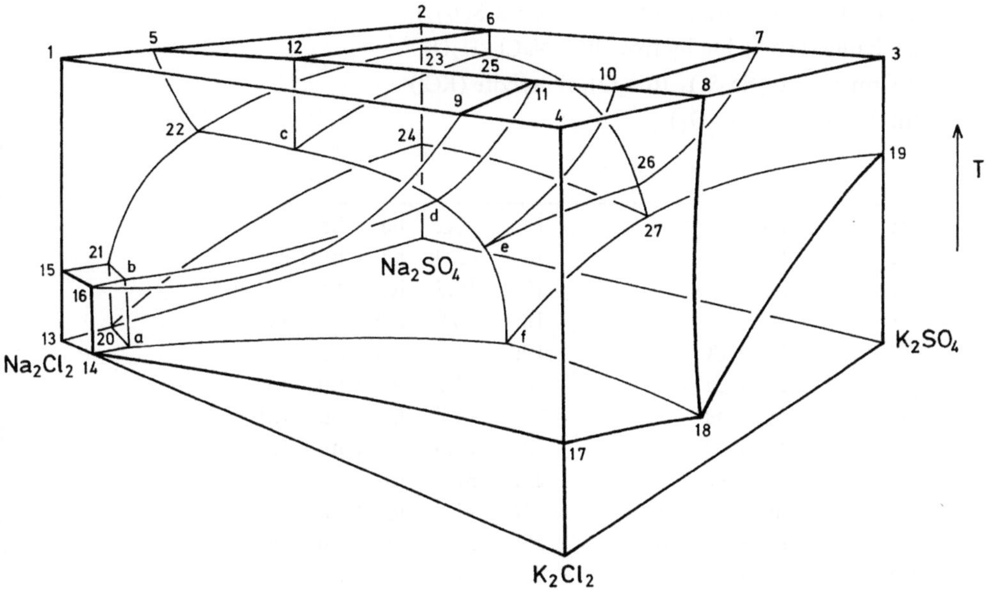

1
2
3
4
5
6
7
8
9
10
11
12
13
14
15
16
17
18
19
20
21
22
23
24
25
26
27
a
b
c
d
e
f
T
Na_2SO_4
K_2SO_4
Na_2Cl_2
K_2Cl_2

III.4.3
Na_2Cl_2-K_2Cl_2-Na_2SO_4-K_2SO_4-H_2O

ac: arcanite (K_2SO_4), gs: glaserite ($K_3Na(SO_4)_2$)
ha: halite (NaCl), hh: hydrohalite ($NaCl{\cdot}2H_2O$)
mi: mirabilite ($Na_2SO_4{\cdot}10H_2O$), sy: sylvite (KCl)
th: thenardite (Na_2SO_4)

	°C			°C	
1:		ha	15:	0.2	ha + hh
2:		th	16:	−2.3	ha + hh + sy
3:		ac	17:	−10.7	ice + sy
4:		sy	18:	−10.9	ice + ac + sy
5:		ha + th	19:	−1.5	ice + ac
6:		gs + th	20:	−21.7	ice + hh + mi
7:		ac + gs	21:	−2.9	ha + hh + mi
8:		ac + sy	22:	17.9	ha + mi + th
9:		ha + sy	23:	32.4	mi + th
10:		ac + gs + sy	24:	−1.3	ice + mi
11:		gs + ha + sy	25:	30.9	gs + mi + th
12:		gs + ha + th	26:	1.8	ac + gs + mi
13:	−21.2	ice + hh	27:	−3.0	ice + ac + mi
14:	−22.9	ice + hh + sy			
a:	−23.3	ice + hh + mi + sy			
b:	−2.4	ha + hh + mi + sy			
c:	16.3	gs + ha + mi + th			
d:	4.4	gs + ha + mi + sy			
e:	−5.8	ac + gs + mi + sy			
f:	−14.0	ice + ac + mi + sy			

SALSYS
Input: temperature. Output: $\{Na_2Cl_2\}$, $\{K_2Cl_2\}$, $\{Na_2SO_4\}$

		°C		
c–12:	gs + ha + th	16.3	-	100
c–d:	gs + ha + mi	4.4	-	16.3
d–11:	gs + ha + sy	4.4	-	100
e–10:	ac + gs + sy	−5.8	-	100

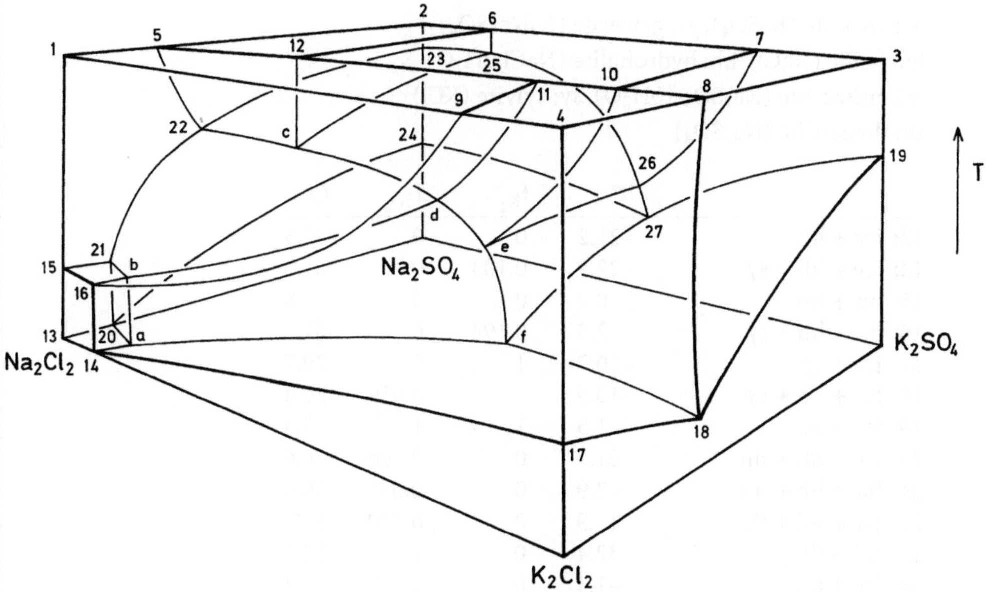
1
2
3
4
5
6
7
8
9
10
11
12
13
14
15
16
17
18
19
20
21
22
23
24
25
26
27
a
b
c
d
e
f
Na_2SO_4
K_2SO_4
Na_2Cl_2
K_2Cl_2
T

III.4.3
Na_2Cl_2-K_2Cl_2-Na_2SO_4-K_2SO_4-H_2O

ac: arcanite (K_2SO_4), gs: glaserite ($K_3Na(SO_4)_2$)
ha: halite (NaCl), hh: hydrohalite ($NaCl\cdot 2H_2O$)
mi: mirabilite ($Na_2SO_4\cdot 10H_2O$), sy: sylvite (KCl)
th: thenardite (Na_2SO_4)

	°C	f_{K_2}	f_{SO_4}	C_T
13: ice + hh	−21.2	0	0	47.3
14: ice + hh + sy	−22.9	0.183	0	51.5
15: ha + hh	0.2	0	0	54.8
16: ha + hh + sy	−2.3	0.194	0	61.2
17: ice + sy	−10.7	1	0	29.7
18: ice + ac + sy	−10.9	1	0.039	30.4
19: ice + ac	−1.5	1	1	7.3
20: ice + hh + mi	−21.7	0	0.006	46.6
21: ha + hh + mi	−2.9	0	0.036	56.0
22: ha + mi + th	17.9	0	0.220	62.8
23: mi + th	32.4	0	1	63.1
24: ice + mi	−1.3	0	1	5.4
25: gs + mi + th	30.9	0.113	1	69.2
26: ac + gs + mi	1.8	0.534	1	17.4
27: ice + ac + mi	−3.0	0.589	1	15.1
a: ice + hh + mi + sy	−23.3	0.150	0.008	52.0
b: ha + hh + mi + sy	−2.4	0.206	0.048	63.0
c: gs + ha + mi + th	16.3	0.125	0.238	68.8
d: gs + ha + mi + sy	4.4	0.218	0.082	66.2
e: ac + gs + mi + sy	−5.8	0.426	0.067	41.8
f: ice + ac + mi + sy	−14.0	0.703	0.041	34.4

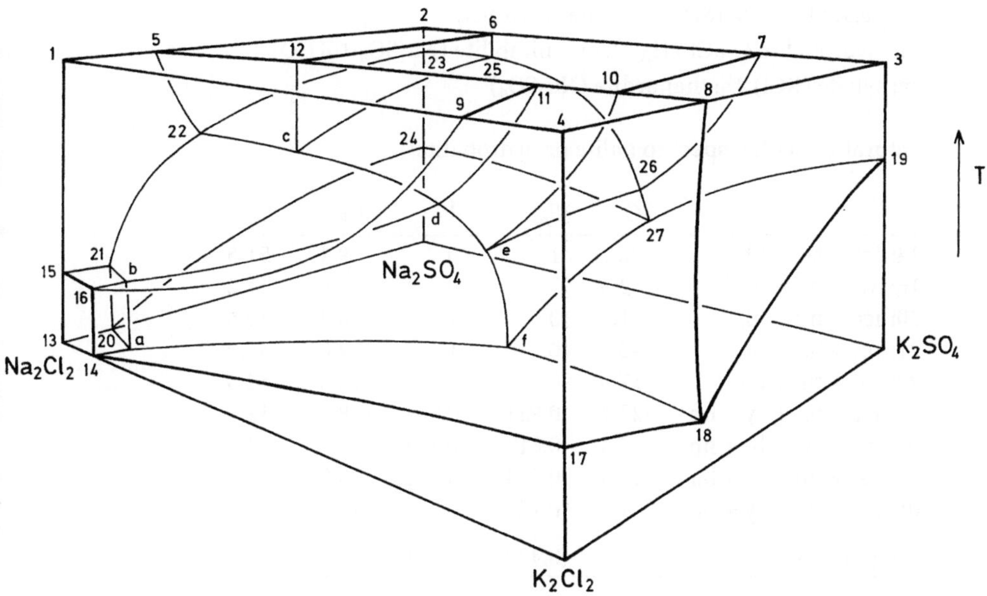

1
2
3
4
5
6
7
8
9
10
11
12
13
14
15
16
17
18
19
20
21
22
23
24
25
26
27
a
b
c
d
e
f
T
Na_2SO_4
K_2SO_4
Na_2Cl_2
K_2Cl_2

III.4.3
Na_2Cl_2-K_2Cl_2-Na_2SO_4-K_2SO_4-H_2O

gs: glaserite ($K_3Na(SO_4)_2$), ha: halite (NaCl)
hh: hydrohalite ($NaCl\cdot 2H_2O$), mi: mirabilite ($Na_2SO_4\cdot 10H_2O$)
sy: sylvite (KCl), th: thenardite (Na_2SO_4)

Saturation with respect to halite or hydrohalite

	°C	F_{K_2}	F_{SO_4}	C_P	C_T
14: ice + sy + hh	−22.9	1	0	9.4	51.5
16: sy + ha + hh	−2.3	1	0	11.9	61.2
20: ice + mi + hh	−21.7	0	1	0.3	46.6
21: mi + ha + hh	−2.9	0	1	2.0	56.0
22: mi + th + ha	17.9	0	1	13.8	62.8
a: ice + mi + sy + hh	−23.3	0.951	0.049	8.2	52.0
b: mi + sy + ha + hh	−2.4	0.812	0.188	16	63
c: gs + mi + th + ha	16.3	0.344	0.656	25.0	68.8
d: gs + mi + sy + ha	4.4	0.727	0.273	19.8	66.2
salt point: gs		0.429	0.571		

$$F_{K_2} = \frac{\{K_2^{2+}\}}{C_P} \qquad F_{SO_4} = \frac{\{SO_4^{2-}\}}{C_P}$$

$$C_P = \{K_2^{2+}\} + \{SO_4^{2-}\}$$

$$C_T = \{Na_2^{2+}\} + \{K_2^{2+}\} = \{Cl_2^{2-}\} + \{SO_4^{2-}\}$$

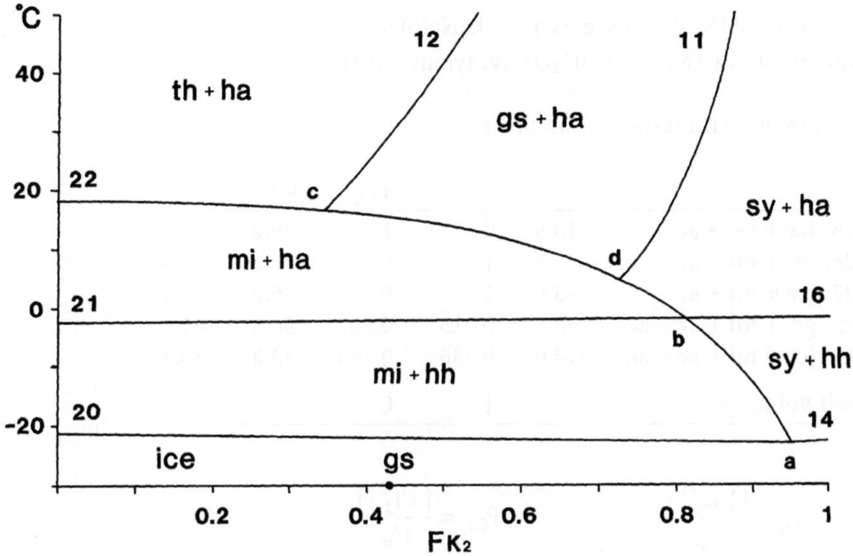

°C
40
20
0
-20
0.2
0.4
0.6
0.8
1
FK2
th + ha
gs + ha
sy + ha
mi + ha
mi + hh
sy + hh
ice
gs
12
11
22
21
20
16
14
c
d
b
a

III.4.3
Na_2Cl_2-K_2Cl_2-Na_2SO_4-K_2SO_4-H_2O

ac: arcanite (K_2SO_4), gs: glaserite ($K_3Na(SO_4)_2$)
mi: mirabilite ($Na_2SO_4 \cdot 10H_2O$), sy: sylvite (KCl)

Saturation with respect to arcanite

	°C	F_{Na_2}	F_{Cl_2}	C_P	C_T
18: ice + sy + ac	−10.9	0	1	29.2	30.4
26: gs + mi + ac	1.8	1	0	8.1	17.4
27: ice + mi + ac	−3.0	1	0	6.2	15.1
e: gs + mi + sy + ac	−5.8	0.313	0.687	56.8	41.8
f: ice + mi + sy + ac	−14.0	0.236	0.764	43.2	34.4
salt point: gs		1	0		

$$F_{Na_2} = \frac{\{Na_2^{2+}\}}{C_P} \qquad F_{Cl_2} = \frac{\{Cl_2^{2-}\}}{C_P}$$

$$C_P = \{Na_2^{2+}\} + \{Cl_2^{2-}\}$$

$$C_T = \{Na_2^{2+}\} + \{K_2^{2+}\} = \{Cl_2^{2-}\} + \{SO_4^{2-}\}$$

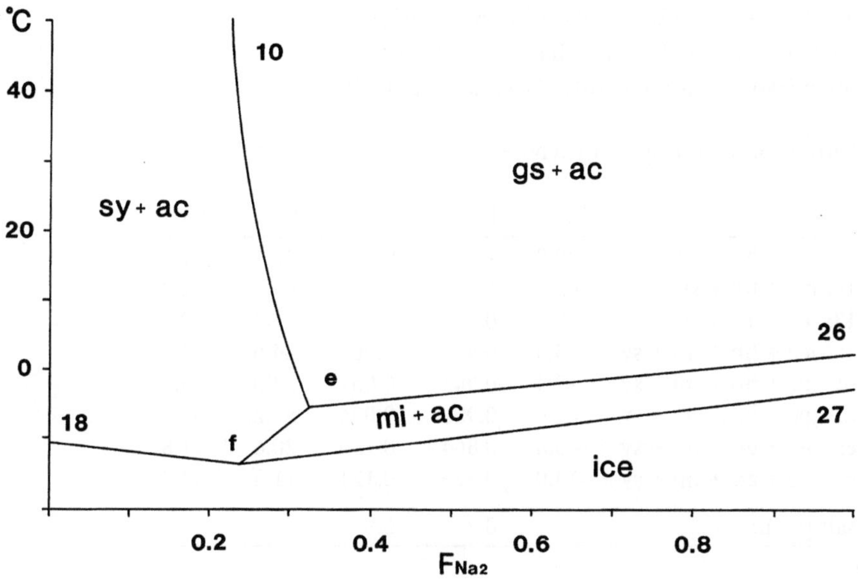

°C
40
20
0
10
gs + ac
sy + ac
26
e
mi + ac
27
18
f
ice
0.2
0.4
0.6
0.8
F_{Na2}

III.4.3
Na_2Cl_2-K_2Cl_2-Na_2SO_4-K_2SO_4-H_2O

ac: arcanite (K_2SO_4), gs: glaserite ($K_3Na(SO_4)_2$)
ha: halite (NaCl), hh: hydrohalite ($NaCl\cdot 2H_2O$)
mi: mirabilite ($Na_2SO_4\cdot 10H_2O$), sy: sylvite (KCl)

Saturation with respect to sylvite

	°C	F_{Na_2}	F_{SO_4}	C_P	C_T
14: ice + hh + sy	−22.9	1	0	42.1	51.5
16: ha + hh + sy	−2.3	1	0	49.3	61.2
18: ice + ac + sy	−10.9	0	1	1.2	30.4
a: ice + hh + mi + sy	−23.3	0.991	0.009	44.6	52.0
b: ha + hh + mi + sy	−2.4	0.943	0.057	53.0	63.0
d: gs + ha + mi + sy	4.4	0.906	0.094	57.2	66.2
e: ac + gs + mi + sy	−5.8	0.864	0.136	20.6	41.8
f: ice + ac + mi + sy	−14.0	0.879	0.121	11.6	34.4
salt point: gs		0.2	0.8		

$$F_{Na_2} = \frac{\{Na_2^{2+}\}}{C_P} \qquad F_{SO_4} = \frac{\{SO_4^{2-}\}}{C_P}$$

$$C_P = \{Na_2^{2+}\} + \{SO_4^{2-}\}$$

$$C_T = \{Na_2^{2+}\} + \{K_2^{2+}\} = \{Cl_2^{2-}\} + \{SO_4^{2-}\}$$

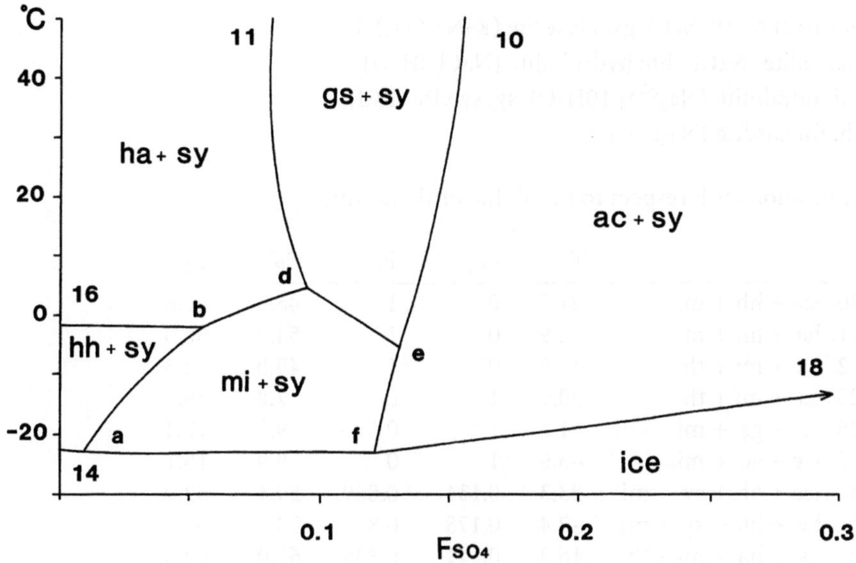

°C
40
20
0
-20
11
10
gs + sy
ha + sy
ac + sy
16
d
b
hh + sy
e
mi + sy
18
a
f
14
ice
0.1
0.2
0.3
FSO4

III.4.3
Na_2Cl_2-K_2Cl_2-Na_2SO_4-K_2SO_4-H_2O

ac: arcanite (K_2SO_4), gs: glaserite ($K_3Na(SO_4)_2$)
ha: halite (NaCl), hh: hydrohalite ($NaCl{\cdot}2H_2O$)
mi: mirabilite ($Na_2SO_4{\cdot}10H_2O$), sy: sylvite (KCl)
th: thenardite (Na_2SO_4)

Saturation with respect to mirabilite or thenardite

	°C	F_{K_2}	F_{Cl_2}	C_P	C_T
20: ice + hh + mi	−21.7	0	1	46.3	46.6
21: ha + hh + mi	−2.9	0	1	54.0	56.0
22: ha + mi + th	17.9	0	1	49.0	62.8
25: gs + mi + th	30.9	1	0	7.8	69.2
26: ac + gs + mi	1.8	1	0	9.3	17.4
27: ice + ac + mi	−3.0	1	0	8.9	15.1
a: ice + hh + sy + mi	−23.3	0.131	0.869	59.4	52.0
b: ha + hh + sy + mi	−2.4	0.178	0.822	73	63.0
c: gs + ha + mi + th	16.3	0.141	0.859	61.0	68.8
d: gs + ha + sy + mi	4.4	0.191	0.809	75.2	66.2
e: ac + gs + sy + mi	−5.8	0.381	0.619	63.0	41.8
f: ice + ac + sy + mi	−14.0	0.423	0.577	57.2	34.4
salt point: gs		1	0		

$$F_{K_2} = \frac{\{K_2^{2+}\}}{C_P} \qquad F_{Cl_2} = \frac{\{Cl_2^{2-}\}}{C_P}$$

$$C_P = \{K_2^{2+}\} + \{Cl_2^{2-}\}$$

$$C_T = \{Na_2^{2+}\} + \{K_2^{2+}\} = \{Cl_2^{2-}\} + \{SO_4^{2-}\}$$

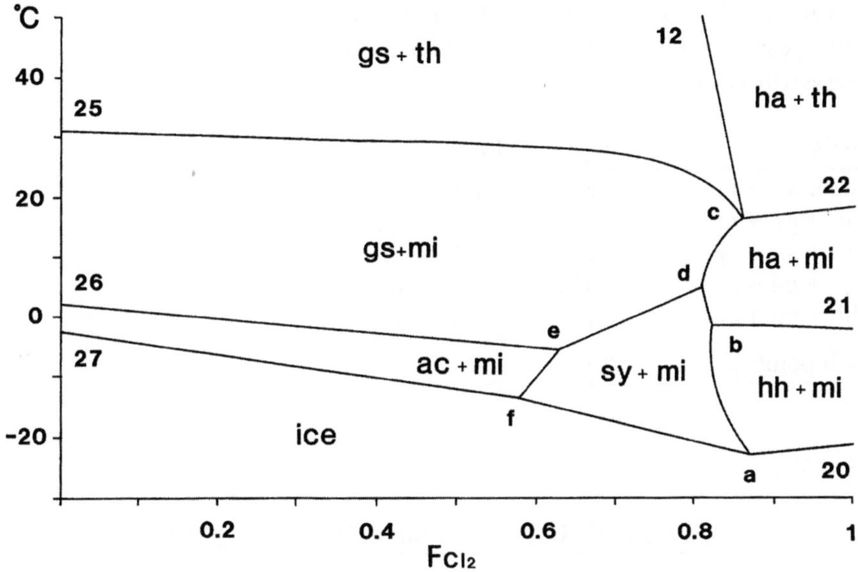

°C
40
20
0
-20
25
26
27
12
22
21
20
gs + th
ha + th
gs+mi
ha + mi
ac + mi
sy + mi
hh + mi
ice
a
b
c
d
e
f
0.2
0.4
0.6
0.8
1
FCl2

III.4.3
Na_2Cl_2-K_2Cl_2-Na_2SO_4-K_2SO_4-H_2O

ac: arcanite (K_2SO_4), gs: glaserite ($K_3Na(SO_4)_2$)
hh: hydrohalite ($NaCl \cdot 2H_2O$), mi: mirabilite ($Na_2SO_4 \cdot 10H_2O$)
sy: sylvite (KCl)

–5°C	f_{K_2}	f_{SO_4}	C_T
ice + ac + mi	0.580	0.436	18.1
ac + gs + mi	0.567	0.236	25.4
ac + gs + sy	0.576	0.067	42.0
gs + mi + sy	0.452	0.071	46.5
hh + mi + sy	0.197	0.031	61.0
salt point: gs	0.75	1	

SALSYS

	Input: $\{K_2Cl_2\}$			Output: $\{Na_2Cl_2\}$, $\{K_2SO_4\}$
	min.		max.	
ice + ac	2.6	–	13.0	
ac + gs	8.4	–	21.4	
ac + mi	2.6	–	8.4	
ac + sy	21.4	–	31.5	
gs + mi	8.4	–	17.7	
gs + sy	17.7	–	21.4	
mi + sy	10.1	–	17.7	

	Input: $\{K_2Cl_2\}$			Output: $\{Na_2Cl_2\}$, $\{Na_2SO_4\}$
	min.		max.	
ice + mi	0	–	10.5	

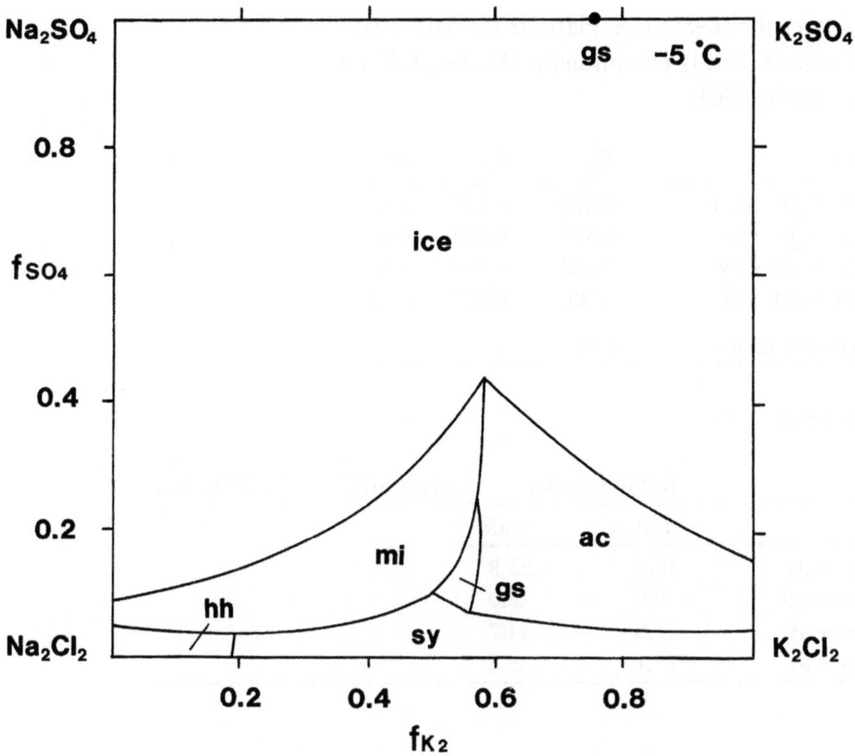

Na2SO4
K2SO4
gs
-5 °C
0.8
fSO4
ice
0.4
0.2
mi
ac
gs
hh
sy
Na2Cl2
K2Cl2
0.2
0.4
0.6
0.8
fK2

III.4.3
Na_2Cl_2-K_2Cl_2-Na_2SO_4-K_2SO_4-H_2O

ac: arcanite (K_2SO_4), gs: glaserite ($K_3Na(SO_4)_2$)
ha: halite (NaCl), mi: mirabilite ($Na_2SO_4 \cdot 10H_2O$)
sy: sylvite (KCl)

0°C	f_{K_2}	f_{SO_4}	C_T
ac + gs + mi	0.478	0.179	33.5
ac + gs + sy	0.593	0.064	43.5
gs + mi + sy	0.262	0.059	57.7
ha + mi + sy	0.206	0.057	64.5
salt point: gs	0.75	1	

SALSYS

	Input: $\{K_2Cl_2\}$			Output: $\{Na_2Cl_2\}$, $\{Na_2SO_4\}$
	min.		max.	
ac + gs	16.0	–	25.8	
ac + mi	9.0	–	16.0	
ac + sy	25.8	–	34.7	
gs + sy	15.1	–	25.8	

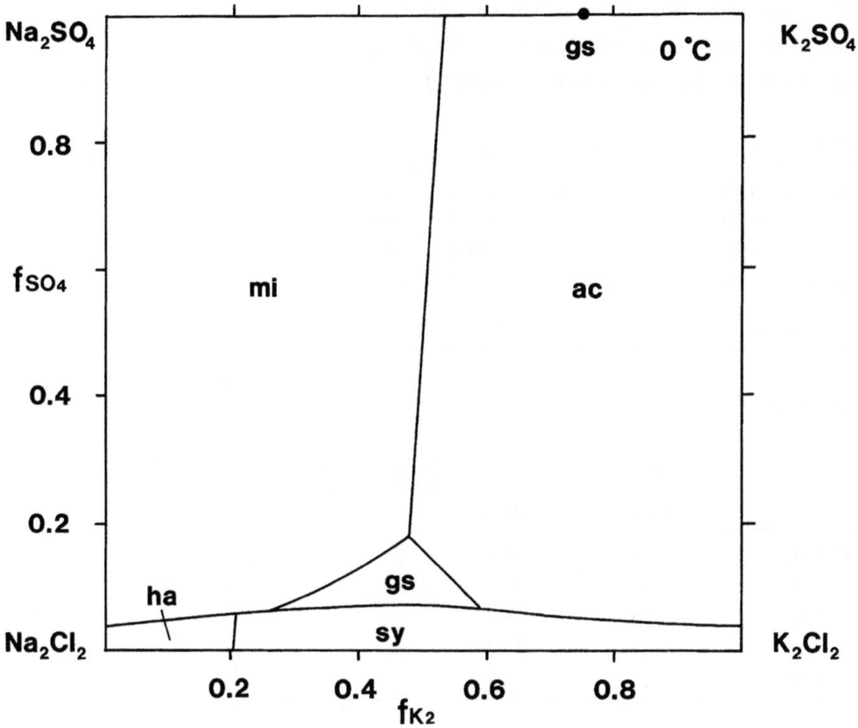

Na_2SO_4
K_2SO_4
gs
0 °C
0.8
f_{SO_4}
mi
ac
0.4
0.2
gs
ha
sy
Na_2Cl_2
K_2Cl_2
0.2
0.4
0.6
0.8
f_{K_2}

III.4.3
Na_2Cl_2-K_2Cl_2-Na_2SO_4-K_2SO_4-H_2O

ac: arcanite (K_2SO_4), gs: glaserite ($K_3Na(SO_4)_2$)
ha: halite (NaCl), mi: mirabilite ($Na_2SO_4 \cdot 10H_2O$)
sy: sylvite (KCl), th: thenardite (Na_2SO_4)

25°C	f_{K_2}	f_{SO_4}	C_T
ac + gs + sy	0.668	0.054	51.5
gs + ha + th	0.143	0.217	69.0
gs + ha + sy	0.291	0.065	69.0
gs + mi + th	0.121	0.585	62.6
salt point: gs	0.75	1	

SALSYS

	Input: $\{K_2Cl_2\}$			Output: $\{Na_2Cl_2\}$, $\{Na_2SO_4\}$
	min.		max.	
ac + sy	34.4	–	44.1	
gs + ha	9.9	–	20.1	
gs + mi	7.6	–	8.7	
gs + sy	20.1	–	34.4	
gs + th	7.6	–	9.9	

	Input: $\{K_2Cl_2\}$			Output: $\{Na_2SO_4\}$, $\{K_2SO_4\}$
	min.		max.	
ac + gs	0	–	48.7	

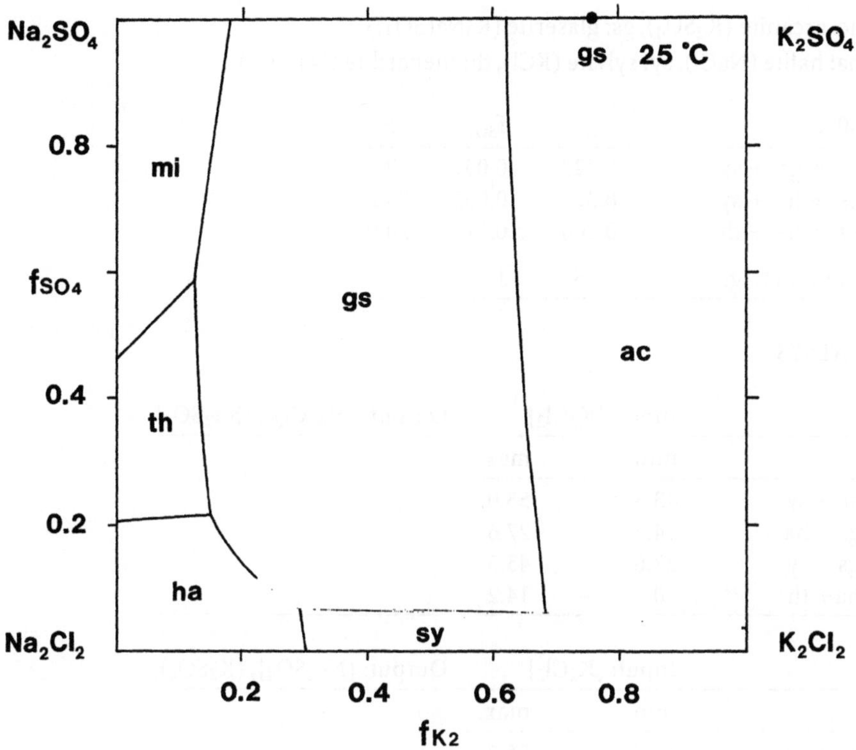

Na2SO4
K2SO4
gs
25 °C
0.8
mi
fSO4
gs
ac
0.4
th
0.2
ha
sy
Na2Cl2
K2Cl2
0.2
0.4
0.6
0.8
fK2

III.4.3 Na_2Cl_2-K_2Cl_2-Na_2SO_4-K_2SO_4-H_2O

ac: arcanite (K_2SO_4), gs: glaserite ($K_3Na(SO_4)_2$)
ha: halite (NaCl), sy: sylvite (KCl), th: thenardite (Na_2SO_4)

50°C	f_{K_2}	f_{SO_4}	C_T
ac + gs + sy	0.727	0.052	59.8
gs + ha + sy	0.377	0.055	73.2
gs + ha + th	0.200	0.172	70.9
salt point: gs	0.75	1	

SALSYS

	Input: {K_2Cl_2}			Output: {Na_2Cl_2}, {Na_2SO_4}
	min.		max.	
ac + sy	43.5	–	53.0	
gs + ha	14.2	–	27.6	
gs + sy	27.6	–	43.5	
ha + th	0	–	14.2	

	Input: {K_2Cl_2}			Output: {Na_2SO_4}, {K_2SO_4}
	min.		max.	
ac + gs	0	–	56.7	

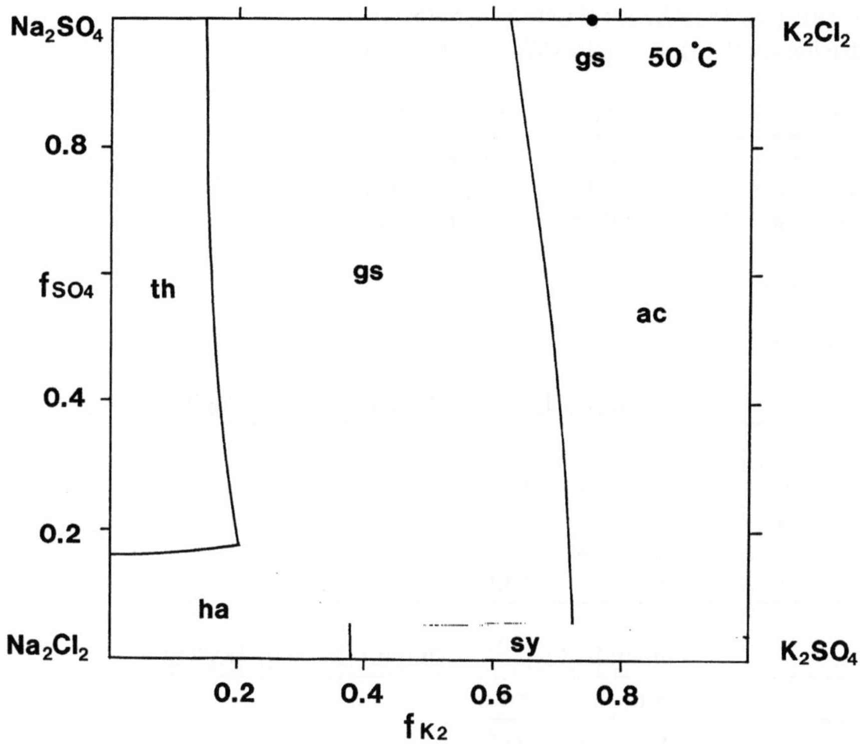

Na2SO4
K2Cl2
gs
50 °C
0.8
fSO4
th
gs
ac
0.4
0.2
ha
Na2Cl2
sy
K2SO4
0.2
0.4
0.6
0.8
fK2

III.4.3
Na_2Cl_2-K_2Cl_2-Na_2SO_4-K_2SO_4-H_2O

ac: arcanite (K_2SO_4), gs: glaserite ($K_3Na(SO_4)_2$)
ha: halite (NaCl), sy: sylvite (KCl), th: thenardite (Na_2SO_4)

75°C	f_{K_2}	f_{SO_4}	C_T
ac + gs + sy	0.777	0.054	68.5
gs + ha + sy	0.447	0.054	79.4
gs + ha + th	0.257	0.150	74.2
salt point: gs	0.75	1	

SALSYS

	Input: $\{K_2Cl_2\}$			Output: $\{Na_2Cl_2\}$, $\{Na_2SO_4\}$
	min.		max.	
ac + sy	53.2	–	61.2	
gs + ha	19.1	–	35.5	
gs + sy	35.5	–	53.2	
ha + th	0	–	19.1	

	Input: $\{K_2Cl_2\}$			Output: $\{Na_2SO_4\}$, $\{K_2SO_4\}$
	min.		max.	
ac + gs	0	–	64.8	

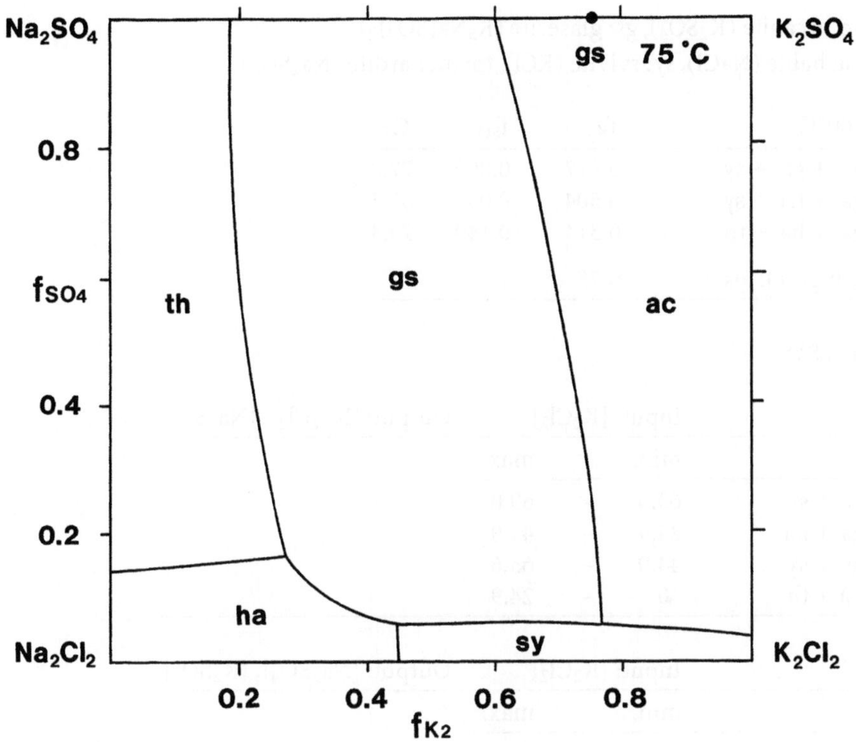

Na_2SO_4
K_2SO_4
gs
75 °C
0.8
f_{SO_4}
th
gs
ac
0.4
0.2
ha
sy
Na_2Cl_2
K_2Cl_2
0.2
0.4
0.6
0.8
f_{K_2}

III.4.3
Na_2Cl_2-K_2Cl_2-Na_2SO_4-K_2SO_4-H_2O

ac: arcanite (K_2SO_4), gs: glaserite ($K_3Na(SO_4)_2$)
ha: halite (NaCl), sy: sylvite (KCl), th: thenardite (Na_2SO_4)

100°C	f_{K_2}	f_{SO_4}	C_T
ac + gs + sy	0.817	0.059	77.8
gs + ha + sy	0.504	0.055	87.3
gs + ha + th	0.314	0.149	79.4
salt point: gs	0.75	1	

SALSYS

	Input: {K_2Cl_2}			Output: {Na_2Cl_2}, {Na_2SO_4}
	min.		max.	
ac + sy	63.6	–	69.0	
gs + ha	24.9	–	44.0	
gs + sy	44.0	–	63.6	
ha + th	0	–	24.9	

	Input: {K_2Cl_2}			Output: {Na_2SO_4}, {K_2SO_4}
	min.		max.	
ac + gs	0	–	73.2	

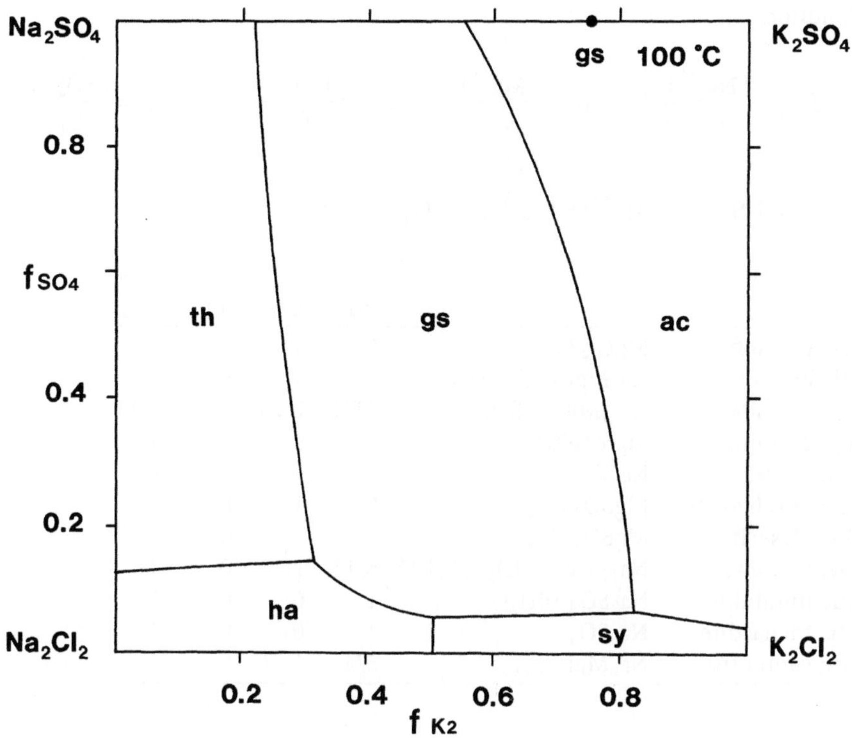

Na_2SO_4
K_2SO_4
gs
100 °C
0.8
f_{SO_4}
th
gs
ac
0.4
0.2
ha
sy
Na_2Cl_2
K_2Cl_2
0.2
0.4
0.6
0.8
f_{K_2}

III.4.4
Na_2Cl_2-$MgCl_2$-Na_2SO_4-$MgSO_4$-H_2O

Solutions

$$f_{Na_2} = \frac{\{Na_2^{2+}\}}{C_T} \qquad f_{Mg} = \frac{\{Mg^{2+}\}}{C_T} \qquad f_{Cl_2} = \frac{\{Cl_2^{2-}\}}{C_T} \qquad f_{SO_4} = \frac{\{SO_4^{2-}\}}{C_T}$$

$$C_T = \{Na_2^{2+}\} + \{Mg^{2+}\} = \{Cl_2^{2-}\} + \{SO_4^{2-}\}$$

Salt points

		f_{Na_2}	f_{Mg}	f_{Cl_2}	f_{SO_4}
bi: bischofite	$MgCl_2{\cdot}6H_2O$	0	1	1	0
bl: bloedite	$Na_2Mg(SO_4)_2{\cdot}4H_2O$	1/2	1/2	0	1
da: d'ansite	$Na_{21}MgCl_3(SO_4)_{10}$	21/23	2/23	3/23	20/23
ep: epsomite	$MgSO_4{\cdot}7H_2O$	0	1	0	1
ha: halite	$NaCl$	1	0	1	0
hx: hexahydrite	$MgSO_4{\cdot}6H_2O$	0	1	0	1
ks: kieserite	$MgSO_4{\cdot}H_2O$	0	1	0	1
lw: loeweite	$Na_{12}Mg_7(SO_4)_{13}{\cdot}15H_2O$	6/13	7/13	0	1
mi: mirabilite	$Na_2SO_4{\cdot}10H_2O$	1	0	0	1
th: thenardite	Na_2SO_4	1	0	0	1
vh: vanthoffite	$Na_6Mg(SO_4)_4$	3/4	1/4	0	1

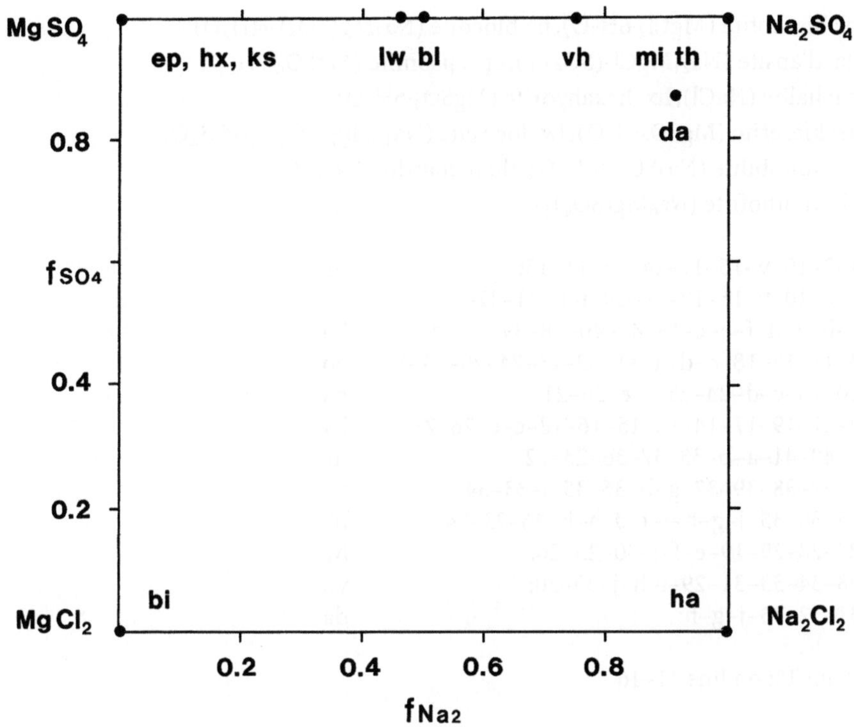
Mg SO4
Na2SO4
ep, hx, ks
lw bl
vh
mi th
da
0.8
fSO4
0.4
0.2
bi
ha
Mg Cl2
Na2Cl2
0.2
0.4
0.6
0.8
fNa2

III.4.4
Na_2Cl_2-$MgCl_2$-Na_2SO_4-$MgSO_4$-H_2O

bi: bischofite ($MgCl_2 \cdot 6H_2O$), bl: bloedite ($Na_2Mg(SO_4)_2 \cdot 4H_2O$)
da: d'ansite ($Na_{21}MgCl_3(SO_4)_{10}$), ep: epsomite ($MgSO_4 \cdot 7H_2O$)
ha: halite (NaCl), hx: hexahydrite ($MgSO_4 \cdot 6H_2O$)
ks: kieserite ($MgSO_4 \cdot H_2O$), lw: loeweite ($Na_{12}Mg_7(SO_4)_{13} \cdot 15H_2O$)
mi: mirabilite ($Na_2SO_4 \cdot 10H_2O$), th: thenardite (Na_2SO_4)
vh: vanthoffite ($Na_6Mg(SO_4)_4$)

1–2–10–9–16–17–14–12–11–15:	bi
5–6–10–9–16–17–19–29–i–h–31–32– g–b–a–d–f–e–c–18–41–40–38–39:	ha
3–11–15–18–c–d–a–41–22–23–24–20–13–12:	ep
20–13–c–d–24–25–f–e–26–21:	hx
4–27–19–17–14–12–15–16–18–c–e–26–21:	ks
7–40–41–a–b–38–37–36–23–22:	mi
8–37–38–39–32–g–b–36–35–j–33–34:	th
25–30–35–j–g–h–i–f–d–a–b–36–23–24:	bl
27–28–29–19–e–f–i–30–25–26:	lw
28–34–33–31–29–i–h–j–35–30:	vh
31–32–33–j–g–h:	da

Point 15: on line 11–16

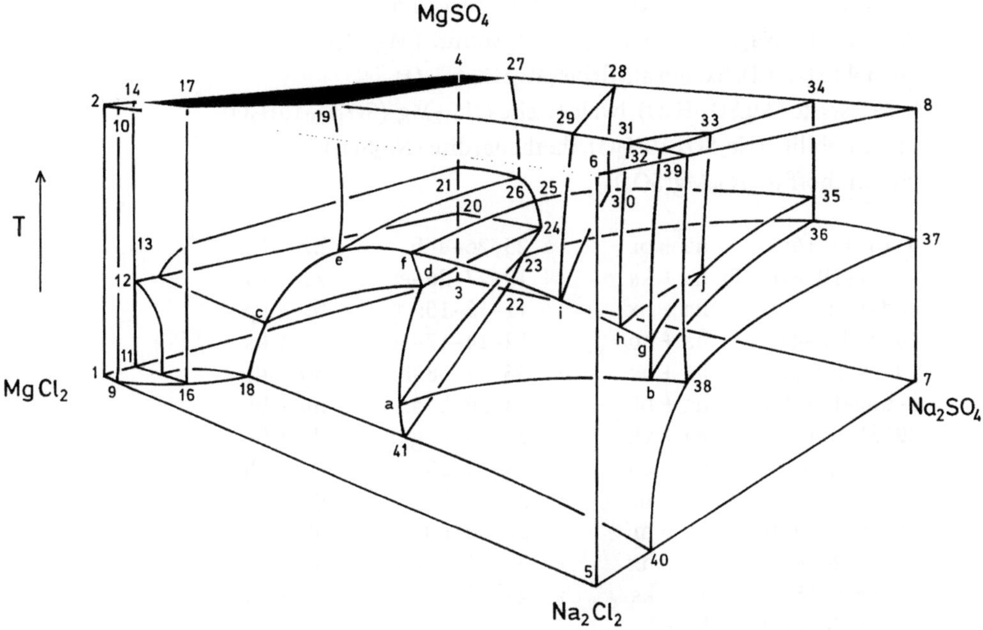

MgSO4
T
MgCl2
Na2Cl2
Na2SO4
1
2
3
4
5
6
7
8
9
10
11
12
13
14
16
17
18
19
20
21
22
23
24
25
26
27
28
29
30
31
32
33
34
35
36
37
38
39
40
41
a
b
c
d
e
f
g
h
i
j

III.4.4
Na_2Cl_2-$MgCl_2$-Na_2SO_4-$MgSO_4$-H_2O

bi: bischofite ($MgCl_2 \cdot 6H_2O$), bl: bloedite ($Na_2Mg(SO_4)_2 \cdot 4H_2O$)
da: d'ansite ($Na_{21}MgCl_3(SO_4)_{10}$), ep: epsomite ($MgSO_4 \cdot 7H_2O$)
ha: halite (NaCl), hx: hexahydrite ($MgSO_4 \cdot 6H_2O$)
ks: kieserite ($MgSO_4 \cdot H_2O$), lw: loeweite ($Na_{12}Mg_7(SO_4)_{13} \cdot 15H_2O$)
mi: mirabilite ($Na_2SO_4 \cdot 10H_2O$), th: thenardite (Na_2SO_4)
vh: vanthoffite ($Na_6Mg(SO_4)_4$)

9–10–17–16:	ha + bi
16–17–19–e–c–18:	ha + ks
c–d–f–e:	ha + hx
18–c–d–a–41:	ha + ep
19–29–i–f–e:	ha + lw
a–b–g–h–i–f–d:	ha + bl
29–31–h–i:	ha + vh
31–32–g–h:	ha + da
38–39–32–g–b:	ha + th
38–40–41–a–b:	ha + mi
13–20–24–d–c:	ep + hx
12–13–c–18–15:	ep + ks
13–21–26–e–c:	hx + ks
22–23–a–41:	ep + mi
23–24–d–a:	ep + bl
24–25–f–d:	hx + bl
25–26–e–f:	hx + lw
19–27–26–e:	ks + lw
11–12–15:	bi + ep
12–14–17–16–15:	bi + ks
36–37–38–b:	mi + th
23–36–b–a:	mi + bl
35–36–b–g–j:	th + bl
33–34–35–j:	th + vh
32–33–j–g:	th + da
25–30–i–f:	bl + lw
30–35–j–h–i:	bl + vh
g–h–j:	bl + da
28–29–i–30:	lw + vh
31–33–j–h:	vh + da

Point 15: on line 11–16

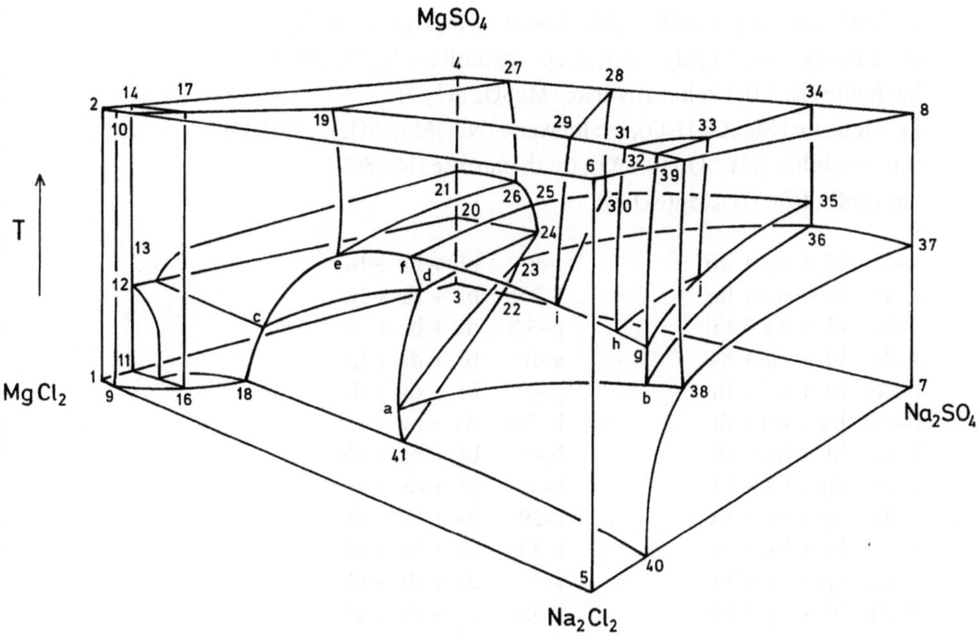
MgSO₄
T
MgCl₂
Na₂SO₄
Na₂Cl₂
1
2
3
4
5
6
7
8
9
10
11
12
13
14
16
17
18
19
20
21
22
23
24
25
26
27
28
29
30
31
32
33
34
35
36
37
38
39
40
41
a
b
c
d
e
f
g
h
i
j

III.4.4
Na_2Cl_2-$MgCl_2$-Na_2SO_4-$MgSO_4$-H_2O

bi: bischofite ($MgCl_2 \cdot 6H_2O$), bl: bloedite ($Na_2Mg(SO_4)_2 \cdot 4H_2O$)
da: d'ansite ($Na_{21}MgCl_3(SO_4)_{10}$), ep: epsomite ($MgSO_4 \cdot 7H_2O$)
ha: halite (NaCl), hx: hexahydrite ($MgSO_4 \cdot 6H_2O$)
ks: kieserite ($MgSO_4 \cdot H_2O$), lw: loeweite ($Na_{12}Mg_7(SO_4)_{13} \cdot 15H_2O$)
mi: mirabilite ($Na_2SO_4 \cdot 10H_2O$), th: thenardite (Na_2SO_4)
vh: vanthoffite ($Na_6Mg(SO_4)_4$)

a–23: bl + ep + mi
a–41: ep + mi + ha
a–b: bl + ha + mi
a–d: bl + ep + ha
b–36: bl + mi + th
b–38: ha + mi + th
b–g: bl + ha + th
c–18: ep + ha + ks
c–d: ep + ha + hx
c–e: ha + hx + ks
c–13: ep + hx + ks
d–24: bl + ep + hx
d–f: bl + ha + hx
e–19: ha + ks + lw
e–26: hx + ks + lw
e–f: ha + hx + lw
f–i: bl + ha + lw
f–25: bl + hx + lw
g–32: da + ha + th
g–h: bl + da + ha
g–j: bl + da + th
h–31: da + ha + vh
h–i: bl + ha + vh
h–j: bl + da + vh
i–29: ha + lw + vh
i–30: bl + lw + vh
j–33: da + th + vh
j–35: bl + th + vh
12–15: bi + ep + ks
16–17: bi + ha + ks

Point 15: on line 11–16

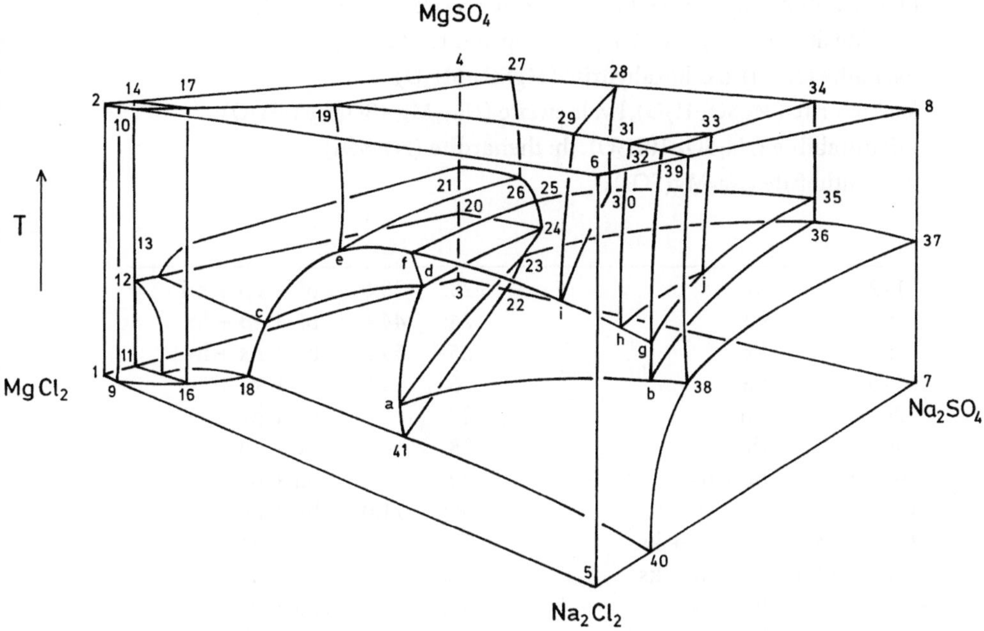
MgSO₄
T
MgCl₂
Na₂SO₄
Na₂Cl₂
1
2
3
4
5
6
7
8
9
10
11
12
13
14
16
17
18
19
20
21
22
23
24
25
26
27
28
29
30
31
32
33
34
35
36
37
38
39
40
41
a
b
c
d
e
f
g
h
i
j

III.4.4
Na_2Cl_2-$MgCl_2$-Na_2SO_4-$MgSO_4$-H_2O

bi: bischofite ($MgCl_2 \cdot 6H_2O$), bl: bloedite ($Na_2Mg(SO_4)_2 \cdot 4H_2O$)
da: d'ansite ($Na_{21}MgCl_3(SO_4)_{10}$), ep: epsomite ($MgSO_4 \cdot 7H_2O$)
ha: halite (NaCl), hx: hexahydrite ($MgSO_4 \cdot 6H_2O$)
ks: kieserite ($MgSO_4 \cdot H_2O$), lw: loeweite ($Na_{12}Mg_7(SO_4)_{13} \cdot 15H_2O$)
mi: mirabilite ($Na_2SO_4 \cdot 10H_2O$), th: thenardite (Na_2SO_4)
vh: vanthoffite ($Na_6Mg(SO_4)_4$)

	°C			°C	
1–2:		bi	23:	20.6	bl + ep + mi
3:		ep	24:	44.4	bl + ep + hx
4:		ks	25:	59.5	bl + hx + lw
5–6:		ha	26:	64.0	hx + ks + lw
7:		mi	27:		ks + lw
8:		th	28:		lw + vh
9–10:		bi + ha	29:		ha + lw + vh
11:		bi + ep	30:	71.0	bl + lw + vh
12:	13.0	bi + ep + ks	31:		da + ha + vh
13:	18.0	ep + hx + ks	32:		da + ha + th
14:		bi + ks	33:		da + th + vh
15:		bi + ep + ks	34:		th + vh
16:		bi + ha + ks	35:	60.0	bl + th + vh
17:		bi + ha + ks	36:	27.0	bl + mi + th
18:		ep + ha + ks	37:	32.4	mi + th
19:		ha + ks + lw	38:	17.9	ha + mi + th
20:	48.2	ep + hx	39:		ha + th
21:	67.5	hx + ks	40:		ha + mi
22:		ep + mi	41:		ep + ha + mi
a:	6.2	bl + ep + ha + mi	g:	45.7	bl + da + ha + th
b:	16.2	bl + ha + mi + th	h:	54.1	bl + da + ha + vh
c:	18.0	ep + ha + hx + ks	i:	61.4	bl + ha + lw + vh
d:	30.2	bl + ep + ha + hx	j:	49	bl + da + th + vh
e:	37.6	ha + hx + ks + lw	l:	–7.8	ep + ha + hh + mi
f:	40.3	bl + ha + hx + lw			

Point 15: on line 11–16, point l below point 41

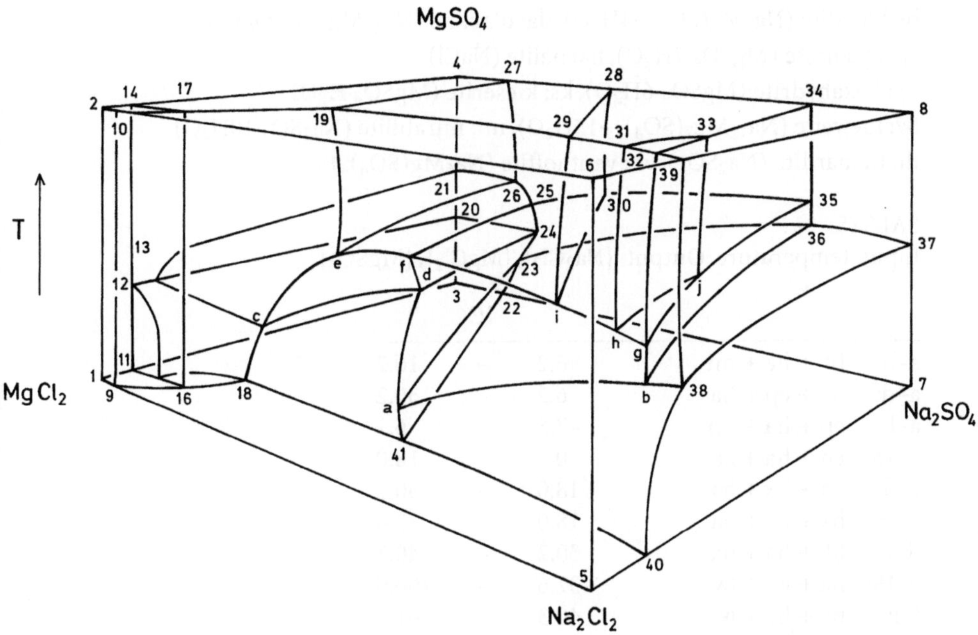
MgSO4
T
MgCl2
Na2SO4
Na2Cl2
a
b
c
d
e
f
g
h
i
j
1
2
3
4
5
6
7
8
9
10
11
12
13
14
16
17
18
19
20
21
22
23
24
25
26
27
28
29
30
31
32
33
34
35
36
37
38
39
40
41

III.4.4
Na_2Cl_2-$MgCl_2$-Na_2SO_4-$MgSO_4$-H_2O

bl: bloedite ($Na_2Mg(SO_4)_2{\cdot}4H_2O$), da: d'ansite ($Na_{21}MgCl_3(SO_4)_{10}$)
ep: epsomite ($MgSO_4{\cdot}7H_2O$), ha: halite (NaCl)
hx: hexahydrite ($MgSO_4{\cdot}6H_2O$), ks: kieserite ($MgSO_4{\cdot}H_2O$)
lw: loeweite ($Na_{12}Mg_7(SO_4)_{13}{\cdot}15H_2O$), mi: mirabilite ($Na_2SO_4{\cdot}10H_2O$)
th: thenardite (Na_2SO_4), vh: vanthoffite ($Na_6Mg(SO_4)_4$)

SALSYS
Input: temperature. Output: {Na_2Cl_2}, {$MgCl_2$}, {$MgSO_4$}

		°C		
a–b:	bl + ha + mi	6.2	–	16.2
a–d:	bl + ep + ha	6.2	–	30.2
a–l:	ep + ha + mi	–7.8	–	6.2
c–18:	ep + ha + ks	0	–	18.0
c–d:	ep + hx + ha	18.0	–	30.2
c–e:	hx + ks + ha	18.0	–	37.6
d–f:	bl + ha + hx	30.2	–	40.3
e–19:	ha + ks + lw	37.6	–	100.0
f–i:	bl + ha + lw	40.3	–	61.4
h–31:	da + ha + vh	54.1	–	90.0
h–i:	bl + ha + vh	54.1	–	61.4
i–29:	ha + lw + vh	61.4	–	100.0
16–17:	bi + ha + ks	0	–	90.0

Input: temperature. Output: {$MgCl_2$}, {Na_2SO_4}, {$MgSO_4$}

		°C		
a–23:	bl + ep + mi	6.2	–	20.6
d–24:	bl + ep + hx	30.2	–	44.4
e–26:	hx + ks + lw	37.6	–	64.0
f–25:	bl + hx + lw	40.3	–	59.5

Input: temperature. Output: {Na_2Cl_2}, {Na_2SO_4}, {$MgSO_4$}

		°C		
b–36:	bl + mi + th	16.2	–	27.0
b–g:	bl + ha + th	16.2	–	45.7
g–32:	da + ha + th	45.7	–	95.0
j–33:	da + th + vh	49.0	–	95.0

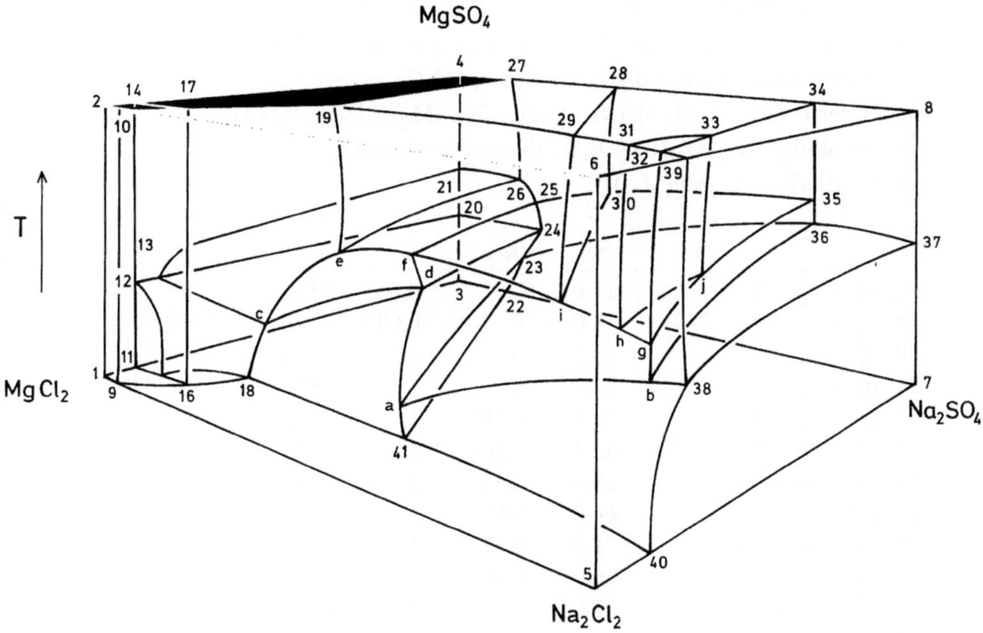
MgSO4
T
MgCl2
Na2Cl2
Na2SO4
1
2
3
4
5
6
7
8
9
10
11
12
13
14
16
17
18
19
20
21
22
23
24
25
26
27
28
29
30
31
32
33
34
35
36
37
38
39
40
41
a
b
c
d
e
f
g
h
i
j

III.4.4
Na_2Cl_2-$MgCl_2$-Na_2SO_4-$MgSO_4$-H_2O

bi: bischofite ($MgCl_2{\cdot}6H_2O$), bl: bloedite ($Na_2Mg(SO_4)_2{\cdot}4H_2O$)
da: d'ansite ($Na_{21}MgCl_3(SO_4)_{10}$), ep: epsomite ($MgSO_4{\cdot}7H_2O$)
ha: halite (NaCl), hx: hexahydrite ($MgSO_4{\cdot}6H_2O$)
ks: kieserite ($MgSO_4{\cdot}H_2O$), lw: loeweite ($Na_{12}Mg_7(SO_4)_{13}{\cdot}15H_2O$)
mi: mirabilite ($Na_2SO_4{\cdot}10H_2O$), th: thenardite (Na_2SO_4)
vh: vanthoffite ($Na_6Mg(SO_4)_4$)

	°C	f_{Na_2}	f_{SO_4}	C_T
12: bi + ep + ks	13.0	0	0.057	105.8
13: ep + hx + ks	18.0	0	0.082	91.0
20: ep + hx	48.2	0	1	73.4
21: hx + ks	67.5	0	1	87.8
23: bl + ep + mi	20.6	0.368	1	73.3
24: bl + ep + hx	44.4	0.167	1	82.1
25: bl + hx + lw	59.5	0.111	1	90.0
26: hx + ks + lw	64.0	0.075	1	90.8
30: bl + lw + vh	71.0	0.510	1	73.5
35: bl + th + vh	60.0	0.598	1	76.1
36: bl + mi + th	27.0	0.614	1	74.3
37: mi + th	32.4	1	1	63.1
38: mi + ha + th	17.9	1	0.220	62.8
a: bl + ep + ha + mi	6.2	0.467	0.280	64.6
b: bl + ha + mi + th	16.2	0.764	0.316	66.4
c: ep + ha + hx + ks	18.0	0.025	0.089	88.7
d: bl + ep + ha + hx	30.2	0.094	0.218	77.0
e: ha + hx + ks + lw	37.6	0.076	0.198	84.4
f: bl + ha + hx + lw	40.3	0.116	0.232	86.1
g: bl + da + ha + th	45.7	0.722	0.259	66.3
h: bl + da + ha + vh	54.1	0.678	0.242	67.3
i: bl + ha + lw + vh	61.4	0.573	0.230	68.8
j: bl + da + th + vh	49	0.646	0.277	66.4
l: ep + ha + hh + mi (not shown)	−7.8	0.331	0.125	61.0

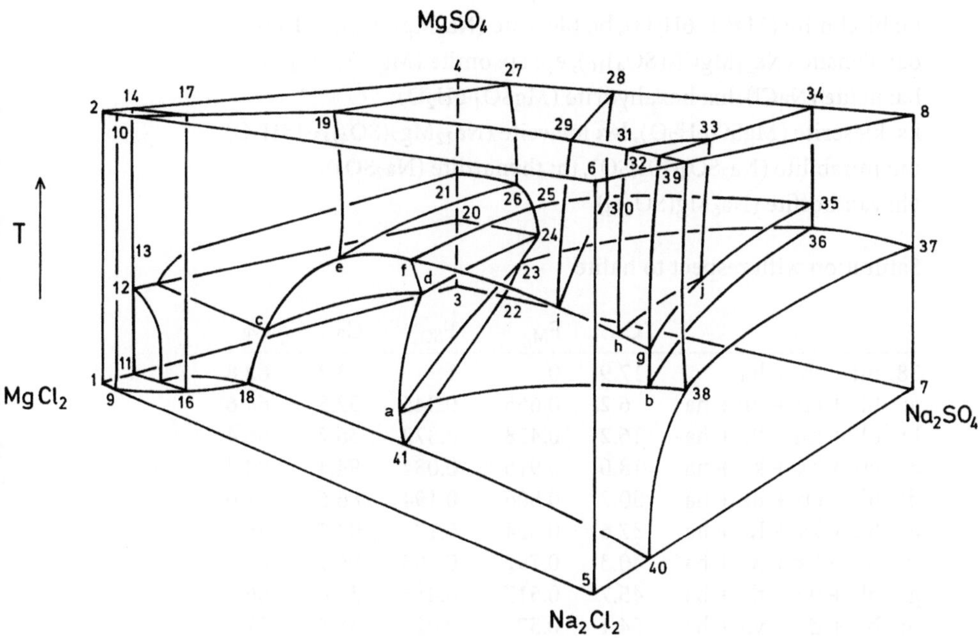
MgSO4
T
MgCl2
Na2Cl2
Na2SO4
1
2
3
4
5
6
7
8
9
10
11
12
13
14
16
17
18
19
20
21
22
23
24
25
26
27
28
29
30
31
32
33
34
35
36
37
38
39
40
41
a
b
c
d
e
f
g
h
i
j

III.4.4
Na_2Cl_2-$MgCl_2$-Na_2SO_4-$MgSO_4$-H_2O

bi: bischofite ($MgCl_2{\cdot}6H_2O$), bl: bloedite ($Na_2Mg(SO_4)_2{\cdot}4H_2O$)
da: d'ansite ($Na_{21}MgCl_3(SO_4)_{10}$), ep: epsomite ($MgSO_4{\cdot}7H_2O$)
ha: halite (NaCl), hx: hexahydrite ($MgSO_4{\cdot}6H_2O$)
ks: kieserite ($MgSO_4{\cdot}H_2O$), lw: loeweite ($Na_{12}Mg_7(SO_4)_{13}{\cdot}15H_2O$)
mi: mirabilite ($Na_2SO_4{\cdot}10H_2O$), th: thenardite (Na_2SO_4)
vh: vanthoffite ($Na_6Mg(SO_4)_4$)

Saturation with respect to halite

	°C	F_{Mg}	F_{SO_4}	C_P	C_T
38: mi + th + ha	17.9	0	1	13.8	62.8
a: bl + ep + mi + ha	6.2	0.655	0.345	52.5	64.6
b: bl + mi + th + ha	16.2	0.428	0.572	36.7	66.4
c: ep + hx + ks + ha	18.0	0.916	0.084	94.4	88.7
d: bl + ep + hx + ha	30.2	0.806	0.194	86.6	77.0
e: hx + ks + lw + ha	37.6	0.824	0.176	94.7	84.4
f: bl + hx + lw + ha	40.3	0.792	0.208	96.1	86.1
g: bl + da + th + ha	45.7	0.517	0.483	35.6	66.3
h: bl + da + vh + ha	54.1	0.571	0.429	38.0	67.3
i: bl + lw + vh + ha	61.4	0.650	0.350	45.2	68.8
l: ep + mi + ha + hh	-7.8	0.843	0.157	48.4	61.0
salt points: bl		0.333	0.667		
da		0.091	0.909		
lw		0.35	0.65		
vh		0.2	0.8		

$$F_{Mg} = \frac{\{Mg^{2+}\}}{C_P} \qquad F_{SO_4} = \frac{\{SO_4^{2-}\}}{C_P}$$

$$C_P = \{Mg^{2+}\} + \{SO_4^{2-}\}$$

$$C_T = \{Na_2^{2+}\} + \{Mg^{2+}\} = \{Cl_2^{2-}\} + \{SO_4^{2-}\}$$

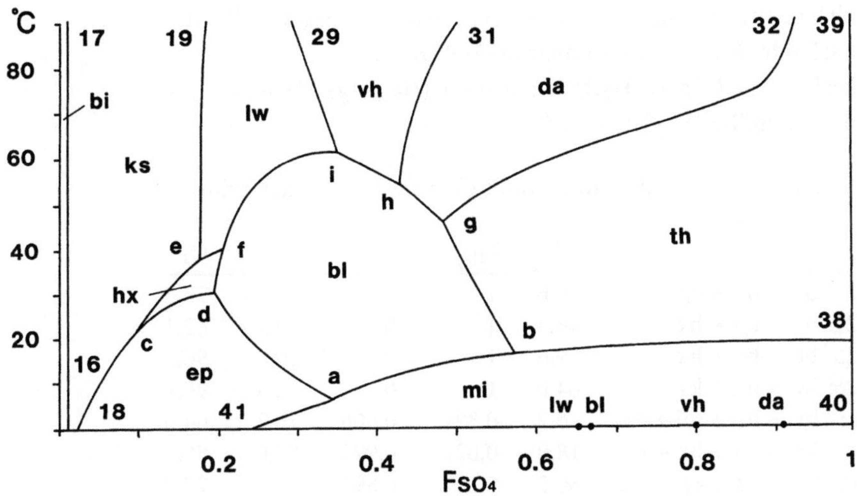
°C
80
60
40
20
17
19
29
31
32
39
bi
ks
lw
vh
da
i
h
g
th
e
f
bl
hx
d
c
16
ep
18
41
a
b
mi
38
lw bl
vh
da
40
0.2
0.4
0.6
0.8
1
FSO4

III.4.4
Na_2Cl_2-$MgCl_2$-Na_2SO_4-$MgSO_4$-H_2O

bl: bloedite ($Na_2Mg(SO_4)_2{\cdot}4H_2O$), ep: epsomite ($MgSO_4{\cdot}7H_2O$)
ha: halite (NaCl), hx: hexahydrite ($MgSO_4{\cdot}6H_2O$)
ks: kieserite ($MgSO_4{\cdot}H_2O$), lw: loeweite ($Na_{12}Mg_7(SO_4)_{13}{\cdot}15H_2O$)
mi: mirabilite ($Na_2SO_4{\cdot}10H_2O$)

Saturation with respect to epsomite, hexahydrite, or kieserite

	°C	F_{Na_2}	F_{Cl_2}	C_P	C_T
23: bl + mi + ep	20.6	1	0	27.0	73.3
24: bl + ep + hx	44.4	1	0	13.7	82.1
25: bl + lw + hx	59.5	1	0	10.0	90.0
26: lw + hx + ks	64.0	1	0	6.8	90.8
a: bl + ha + mi + ep	6.2	0.394	0.606	76.7	64.6
c: ha + ep + hx + ks	18.0	0.027	0.973	83.0	88.7
d: bl + ha + ep + hx	30.2	0.107	0.893	67.4	77.0
e: ha + lw + hx + ks	37.6	0.086	0.914	74.1	84.4
f: bl + ha + lw + hx	40.3	0.131	0.869	76.1	86.1
salt points: bl, lw		1	0		

$$F_{Na_2} = \frac{\{Na_2^{2+}\}}{C_P} \qquad F_{Cl_2} = \frac{\{Cl_2^{2-}\}}{C_P}$$

$$C_P = \{Na_2^{2+}\} + \{Cl_2^{2-}\}$$

$$C_T = \{Na_2^{2+}\} + \{Mg^{2+}\} = \{Cl_2^{2-}\} + \{SO_4^{2-}\}$$

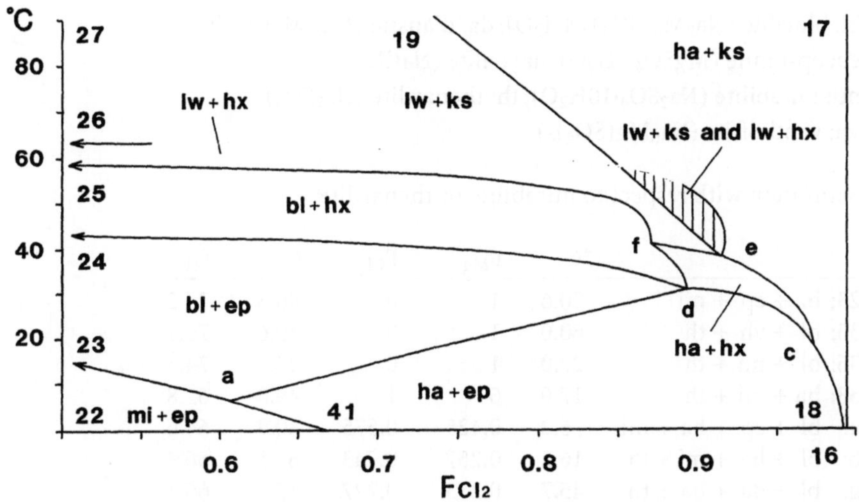

°C
80
60
40
20
27
26
25
24
23
22
19
17
18
16
41
lw + hx
lw + ks
ha + ks
lw + ks and lw + hx
bl + hx
bl + ep
ha + ep
ha + hx
mi + ep
a
c
d
e
f
0.6
0.7
0.8
0.9
FCl2

III.4.4
Na_2Cl_2-$MgCl_2$-Na_2SO_4-$MgSO_4$-H_2O

bl: bloedite ($Na_2Mg(SO_4)_2 \cdot 4H_2O$), da: d'ansite ($Na_{21}MgCl_3(SO_4)_{10}$)
ep: epsomite ($MgSO_4 \cdot 7H_2O$), ha: halite (NaCl)
mi: mirabilite ($Na_2SO_4 \cdot 10H_2O$), th: thenardite (Na_2SO_4)
vh: vanthoffite ($Na_6Mg(SO_4)_4$)

Saturation with respect to mirabilite or thenardite

	°C	F_{Mg}	F_{Cl_2}	C_P	C_T
23: bl + ep + mi	20.6	1	0	46.3	73.3
35: bl + vh + th	60.0	1	0	30.6	76.1
36: bl + mi + th	27.0	1	0	28.7	74.3
38: ha + mi + th	17.9	0	1	49.0	62.8
a: bl + ep + ha + mi	6.2	0.425	0.575	80.9	64.6
b: bl + ha + mi + th	16.2	0.257	0.743	61.1	66.4
g: bl + da + ha + th	45.7	0.273	0.727	67.5	66.3
j: bl + da + vh + th	49	0.329	0.671	71.5	66.4
l: ep + ha + hh + mi	−7.8	0.433	0.567	94.2	61.0
salt points: bl, vh		1	0		
da		0.4	0.6		

$$F_{Mg} = \frac{\{Mg^{2+}\}}{C_P} \qquad F_{Cl_2} = \frac{\{Cl_2^{2-}\}}{C_P}$$

$$C_P = \{Mg^{2+}\} + \{Cl_2^{2-}\}$$

$$C_T = \{Na_2^{2+}\} + \{Mg^{2+}\} = \{Cl_2^{2-}\} + \{SO_4^{2-}\}$$

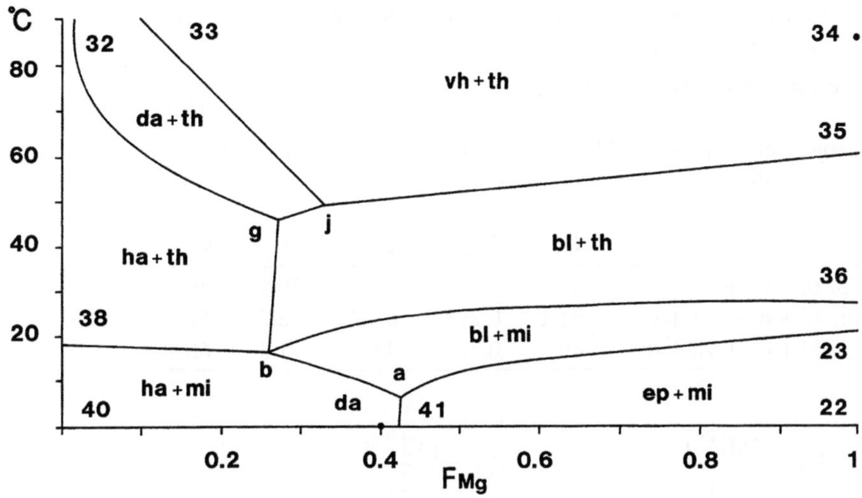

°C
80
60
40
20
32
33
34
35
36
38
23
40
22
41
vh + th
da + th
ha + th
bl + th
bl + mi
ha + mi
da
ep + mi
g
j
b
a
0.2
0.4
0.6
0.8
1
FMg

III.4.4
Na_2Cl_2-$MgCl_2$-Na_2SO_4-$MgSO_4$-H_2O

bl: bloedite ($Na_2Mg(SO_4)_2{\cdot}4H_2O$) da: d'ansite ($Na_{21}MgCl_3(SO_4)_{10}$)
ha: halite (NaCl), th: thenardite (Na_2SO_4)
vh: vanthoffite ($Na_6Mg(SO_4)_4$)

Projection of vh + da (31–33–h–j) and vh + th (33–34–35–j) from Na_2SO_4

	°C	F_{Mg}	F_{Cl_2}	C_P	C_T
35: bl + vh + th	60.0	1	0	30.6	76.1
h: bl + da + ha + vh	54.1	0.298	0.702	72.7	67.3
j: bl + da + vh + th	49	0.329	0.671	71.5	66.4

$$F_{Mg} = \frac{\{Mg^{2+}\}}{C_P} \qquad F_{Cl_2} = \frac{\{Cl_2^{2-}\}}{C_P}$$

$$C_P = \{Mg^{2+}\} + \{Cl_2^{2-}\}$$

$$C_T = \{Na_2^{2+}\} + \{Mg^{2+}\} = \{Cl_2^{2-}\} + \{SO_4^{2-}\}$$

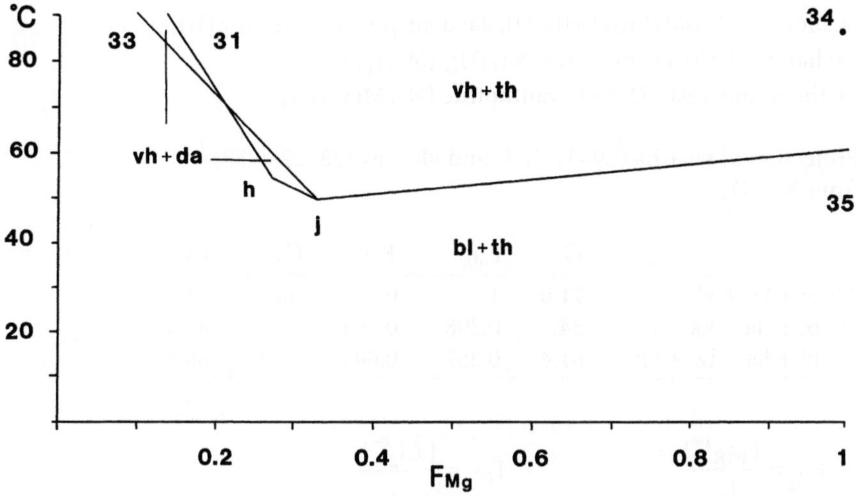
°C
80
60
40
20
33
31
34
vh + th
vh + da
h
j
35
bl + th
0.2
0.4
0.6
0.8
1
F_{Mg}

III.4.4
Na_2Cl_2-$MgCl_2$-Na_2SO_4-$MgSO_4$-H_2O

bl: bloedite ($Na_2Mg(SO_4)_2 \cdot 4H_2O$), da: d'ansite ($Na_{21}MgCl_3(SO_4)_{10}$)
ha: halite (NaCl), lw: loeweite ($Na_{12}Mg_7(SO_4)_{13} \cdot 15H_2O$)
th: thenardite (Na_2SO_4), vh: vanthoffite ($Na_6Mg(SO_4)_4$)

Projection of vh + ha (29–31–h–i) and vh + lw (28–29–i–30) from Na_2SO_4

	°C	F_{Mg}	F_{Cl_2}	C_P	C_T
30: bl + lw + vh	71.0	1	0	36.0	73.5
h: bl + da + ha + vh	54.1	0.298	0.702	72.7	67.3
i: bl + ha + lw + vh	61.4	0.357	0.643	82.4	68.8

$$F_{Mg} = \frac{\{Mg^{2+}\}}{C_P} \qquad F_{Cl_2} = \frac{\{Cl_2^{2-}\}}{C_P}$$

$$C_P = \{Mg^{2+}\} + \{Cl_2^{2-}\}$$

$$C_T = \{Na_2^{2+}\} + \{Mg^{2+}\} = \{Cl_2^{2-}\} + \{SO_4^{2-}\}$$

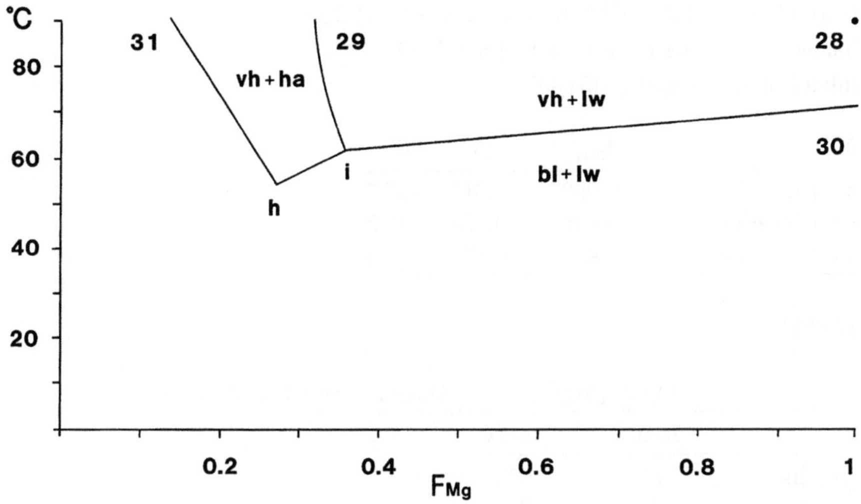

°C
80
60
40
20
31
29
28
30
vh + ha
vh + lw
bl + lw
h
i
0.2
0.4
0.6
0.8
1
F_{Mg}

III.4.4
Na_2Cl_2-$MgCl_2$-Na_2SO_4-$MgSO_4$-H_2O

bi: bischofite ($MgCl_2 \cdot 6H_2O$), ep: epsomite ($MgSO_4 \cdot 7H_2O$)
ha: halite (NaCl), ks: kieserite ($MgSO_4 \cdot H_2O$)
mi: mirabilite ($Na_2SO_4 \cdot 10H_2O$)

0°C	f_{Na_2}	f_{SO_4}	C_T
bi + ha + ks	0.060	0.017	107.7
ep + ha + ks	0.009	0.026	96.9
ep + ha + mi	0.407	0.194	62.4

SALSYS

	Input: {$MgCl_2$}			Output: {Na_2Cl_2}, {$MgSO_4$}
	min.		max.	
ep + ha	24.9	–	93.5	
ep + mi	1.1	–	24.9	

	Input: {$MgCl_2$}			Output: {Na_2Cl_2}, {Na_2SO_4}
	min.		max.	
ha + mi	0	–	37.0	

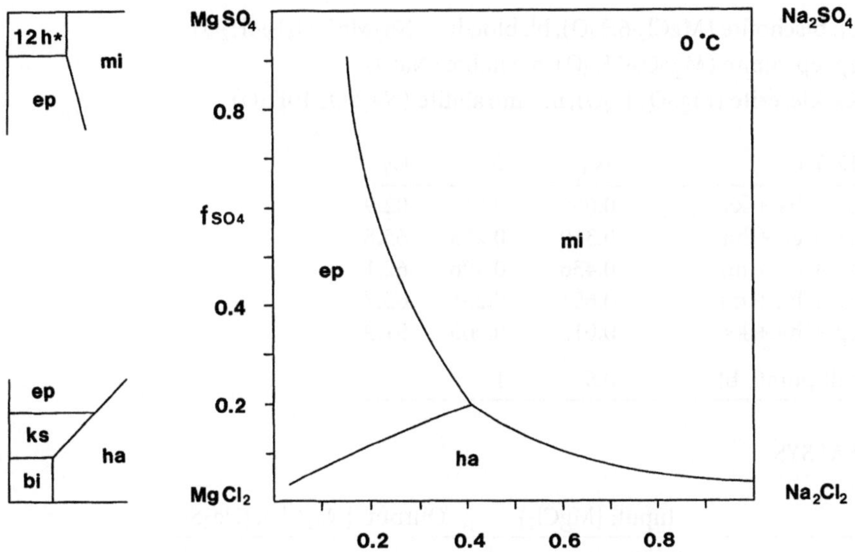
12 h*
mi
ep
ep
ks
ha
bi
Mg SO$_4$
Na$_2$SO$_4$
0 °C
0.8
f SO4
mi
ep
0.4
0.2
ha
Mg Cl$_2$
Na$_2$Cl$_2$
0.2
0.4
0.6
0.8

III.4.4
Na_2Cl_2-$MgCl_2$-Na_2SO_4-$MgSO_4$-H_2O

bi: bischofite ($MgCl_2 \cdot 6H_2O$), bl: bloedite ($Na_2Mg(SO_4)_2 \cdot 4H_2O$)
ep: epsomite ($MgSO_4 \cdot 7H_2O$), ha: halite (NaCl)
ks: kieserite ($MgSO_4 \cdot H_2O$), mi: mirabilite ($Na_2SO_4 \cdot 10H_2O$)

12.5°C	f_{Na_2}	f_{SO_4}	C_T
bi + ha + ks	0.006	0.013	102.9
bl + ep + ha	0.360	0.255	65.8
bl + ep + mi	0.436	0.496	62.1
bl + ha + mi	0.651	0.291	65.7
ep + ha + ks	0.017	0.065	91.9
salt point: bl	0.5	1	

SALSYS

	Input: {$MgCl_2$}			Output: {Na_2Cl_2}, {Na_2SO_4}
	min.		max.	
bl + ha	22.9	–	42.1	
ha + mi	0	–	22.9	

	Input: {$MgCl_2$}			Output: {Na_2Cl_2}, {$MgSO_4$}
	min.		max.	
ep + ha	25.3	–	84.3	

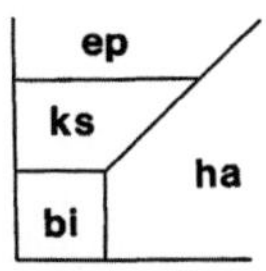
ep
ks
ha
bi

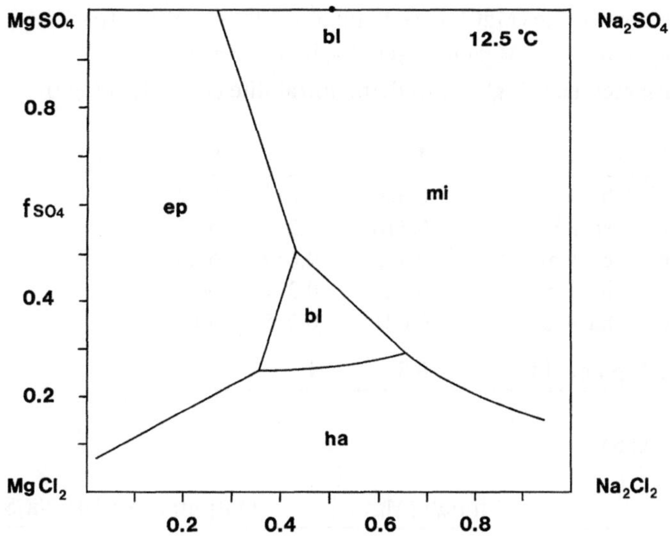
MgSO4
Na2SO4
MgCl2
Na2Cl2
12.5 °C
bl
ep
mi
bl
ha
fSO4
0.8
0.4
0.2
0.2
0.4
0.6
0.8

III.4.4
Na_2Cl_2-$MgCl_2$-Na_2SO_4-$MgSO_4$-H_2O

bi: bischofite ($MgCl_2{\cdot}6H_2O$), bl: bloedite ($Na_2Mg(SO_4)_2{\cdot}4H_2O$)
ep: epsomite ($MgSO_4{\cdot}7H_2O$), ha: halite (NaCl)
ks: kieserite ($MgSO_4{\cdot}H_2O$), mi: mirabilite ($Na_2SO_4{\cdot}10H_2O$)

15°C	f_{Na_2}	f_{SO_4}	C_T
bi + ha + ks	0.006	0.012	103.2
bl + ep + ha	0.318	0.247	66.7
bl + ep + mi	0.417	0.639	63.8
bl + ha + mi	0.727	0.307	66.2
ep + ha + ks	0.021	0.075	90.5
salt point: bl	0.5	1	

SALSYS

	Input: {$MgCl_2$}			Output: {Na_2Cl_2}, {Na_2SO_4}
	min.		max.	
bl + ha	18.1	–	45.5	
ha + mi	0	–	18.1	

	Input: {$MgCl_2$}			Output: {Na_2Cl_2}, {$MgSO_4$}
	min.		max.	
bl + ep	–3.6	–	29.0	
ep + ha	29.0	–	81.8	

	Input: {$MgCl_2$}			Output: {Na_2SO_4}, {$MgSO_4$}
	min.		max.	
ep + mi	0	–	23.0	

	Input: {$MgSO_4$}			Output: {Na_2Cl_2}, {Na_2SO_4}
	min.		max.	
bl + mi	18.1	–	37.2	

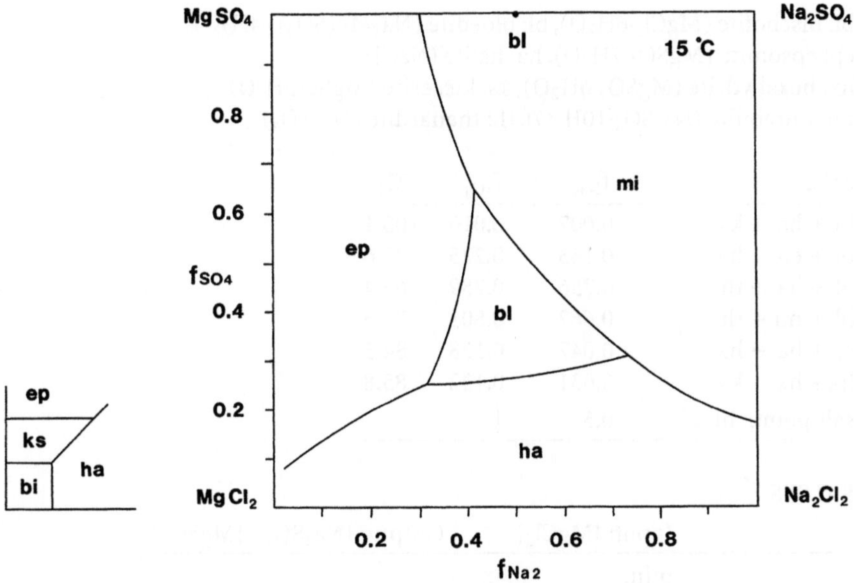

MgSO4
Na2SO4
MgCl2
Na2Cl2
15 °C
bl
mi
ep
bl
ha
fSO4
0.8
0.6
0.4
0.2
0.2
0.4
0.6
0.8
fNa2
ep
ks
bi
ha

III.4.4
Na_2Cl_2-$MgCl_2$-Na_2SO_4-$MgSO_4$-H_2O

bi: bischofite ($MgCl_2 \cdot 6H_2O$), bl: bloedite ($Na_2Mg(SO_4)_2 \cdot 4H_2O$)
ep: epsomite ($MgSO_4 \cdot 7H_2O$), ha: halite (NaCl)
hx: hexahydrite ($MgSO_4 \cdot 6H_2O$), ks: kieserite ($MgSO_4 \cdot H_2O$)
mi: mirabilite ($Na_2SO_4 \cdot 10H_2O$), th: thenardite (Na_2SO_4)

25°C	f_{Na_2}	f_{SO_4}	C_T
bi + ha + ks	0.007	0.010	105.1
bl + ep + ha	0.163	0.225	72.6
bl + ha + th	0.756	0.289	66.4
bl + mi + th	0.662	0.801	71.5
ep + ha + hx	0.047	0.128	84.5
ha + hx + ks	0.031	0.122	85.8
salt point: bl	0.5	1	

SALSYS

	Input: {$MgCl_2$}			Output: {Na_2SO_4}, {$MgSO_4$}
	min.		max.	
bl + ep	0	–	56.3	

	Input: {$MgCl_2$}			Output: {Na_2Cl_2}, {Na_2SO_4}
	min.		max.	
bl + ha	16.2	–	60.8	

	Input: {$MgCl_2$}			Output: {Na_2Cl_2}, {$MgSO_4$}
	min.		max.	
ep + ha	44.5	–	69.7	
ha + ks	72.6	–	103.4	

	Input: {$MgSO_4$}			Output: {Na_2Cl_2}, {Na_2SO_4}
	min.		max.	
bl + mi	24.2	–	36.1	
ha + th	0	–	16.2	
mi + th	0	–	24.2	

	Input: {Na_2Cl_2}			Output: {Na_2SO_4}, {$MgSO_4$}
	min.		max.	
bl + th	14.2	–	47.2	

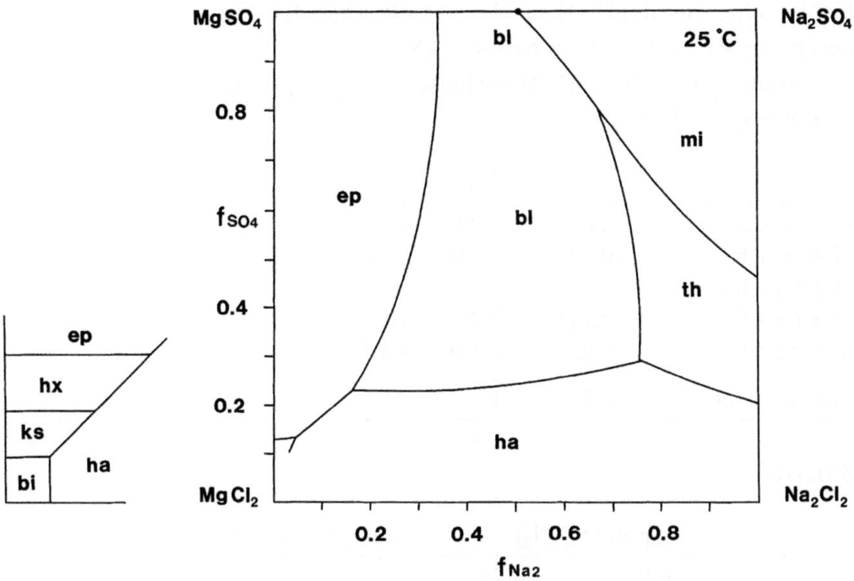
ep
hx
ks
bi
ha
MgSO₄
Na₂SO₄
25 °C
bl
mi
ep
bl
th
ha
0.8
f SO4
0.4
0.2
MgCl₂
Na₂Cl₂
0.2
0.4
0.6
0.8
f Na2

III.4.4
Na_2Cl_2-$MgCl_2$-Na_2SO_4-$MgSO_4$-H_2O

bi: bischofite ($MgCl_2 \cdot 6H_2O$), bl: bloedite ($Na_2Mg(SO_4)_2 \cdot 4H_2O$)
ep: epsomite ($MgSO_4 \cdot 7H_2O$), ha: halite (NaCl)
hx: hexahydrite ($MgSO_4 \cdot 6H_2O$), ks: kieserite ($MgSO_4 \cdot H_2O$)
th: thenardite (Na_2SO_4)

35°C	f_{Na_2}	f_{SO_4}	C_T
bi + ha + ks	0.007	0.007	107.3
bl + ep + hx	0.125	0.326	76.0
bl + ha + hx	0.094	0.227	81.2
bl + ha + th	0.744	0.268	66.4
ha + hx + ks	0.063	0.181	84.3
salt point: bl	0.5	1	

SALSYS

	Input: {$MgCl_2$}			Output: {Na_2Cl_2}, {Na_2SO_4}
	min.		max.	
bl + ha	17.0	–	73.6	

	Input: {$MgCl_2$}			Output: {Na_2SO_4}, {$MgSO_4$}
	min.		max.	
bl + ep	0	–	51.2	

	Input: {$MgSO_4$}			Output: {Na_2Cl_2}, {Na_2SO_4}
	min.		max.	
ha + th	0	–	17.0	

	Input: {Na_2Cl_2}			Output: {Na_2SO_4}, {$MgSO_4$}
	min.		max.	
bl + th	0	–	48.6	

	Input: {Na_2Cl_2}			Output: {$MgCl_2$}, {$MgSO_4$}
	min.		max.	
ha + ks	0.7	–	5.3	

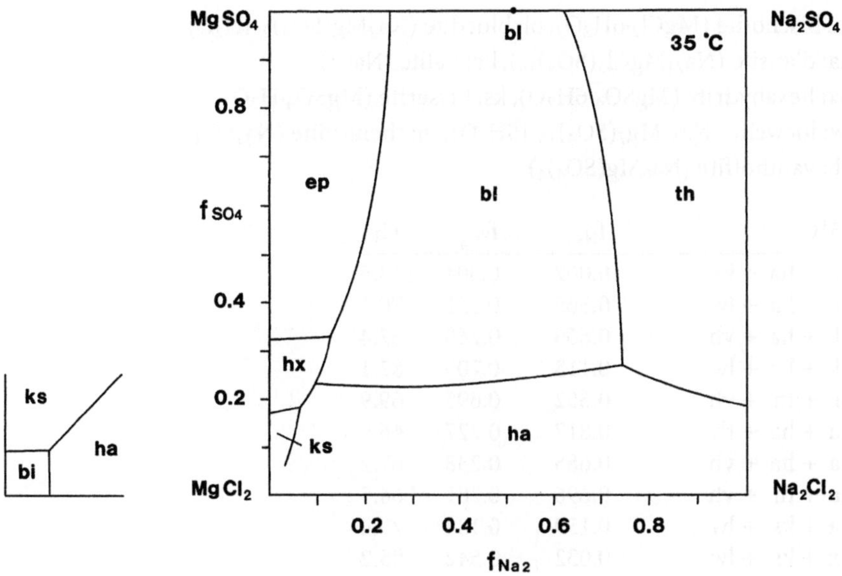
ks
ha
bi
MgSO4
Na2SO4
bl
35 °C
0.8
ep
bl
th
fSO4
0.4
hx
0.2
ha
ks
MgCl2
Na2Cl2
0.2
0.4
0.6
0.8
fNa2

III.4.4
Na_2Cl_2-$MgCl_2$-Na_2SO_4-$MgSO_4$-H_2O

bi: bischofite ($MgCl_2 \cdot 6H_2O$), bl: bloedite ($Na_2Mg(SO_4)_2 \cdot 4H_2O$)
da: d'ansite ($Na_{21}MgCl_3(SO_4)_{10}$), ha: halite (NaCl)
hx: hexahydrite ($MgSO_4 \cdot 6H_2O$), ks: kieserite ($MgSO_4 \cdot H_2O$)
lw: loeweite ($Na_{12}Mg_7(SO_4)_{13} \cdot 15H_2O$), th: thenardite ($Na_2SO_4$)
vh: vanthoffite ($Na_6Mg(SO_4)_4$)

55°C		f_{Na_2}	f_{SO_4}	C_T
bi + ha + ks		0.007	0.004	113.5
bl + ha + lw		0.365	0.221	70.1
bl + ha + vh		0.666	0.240	67.4
bl + hx + lw		0.113	0.705	87.1
bl + th + vh		0.652	0.695	69.9
da + ha + th		0.817	0.227	66.6
da + ha + vh		0.685	0.238	67.2
da + th + vh		0.696	0.281	64.5
ha + ks + lw		0.139	0.184	79.4
hx + ks + lw		0.032	0.542	85.2
salt points:	bl	0.5	1	
	da	0.913	0.870	
	lw	0.462	1	
	vh	0.75	1	

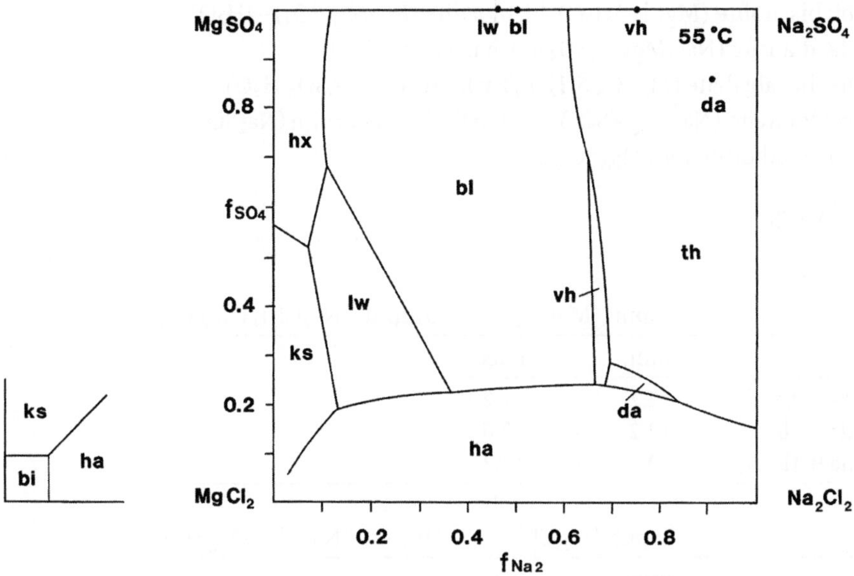

MgSO4
Na2SO4
MgCl2
Na2Cl2
55 °C
lw bl
vh
da
hx
bl
lw
ks
vh
th
da
ha
ks
bi
ha
0.8
0.4
0.2
fSO4
0.2
0.4
0.6
0.8
fNa2

III.4.4
Na_2Cl_2-$MgCl_2$-Na_2SO_4-$MgSO_4$-H_2O

bi: bischofite ($MgCl_2 \cdot 6H_2O$), bl: bloedite ($Na_2Mg(SO_4)_2 \cdot 4H_2O$)
da: d'ansite ($Na_{21}MgCl_3(SO_4)_{10}$), ha: halite (NaCl)
hx: hexahydrite ($MgSO_4 \cdot 6H_2O$), ks: kieserite ($MgSO_4 \cdot H_2O$)
lw: loeweite ($Na_{12}Mg_7(SO_4)_{13} \cdot 15H_2O$), th: thenardite ($Na_2SO_4$)
vh: vanthoffite ($Na_6Mg(SO_4)_4$)

SALSYS

55°C

	Input: {$MgCl_2$}			Output: {Na_2Cl_2}, {Na_2SO_4}
	min.		max.	
da + ha	12.2	-	21.2	
da + th	12.2	-	19.6	
ha + th	0	-	12.2	

	Input: {$MgCl_2$}			Output: {Na_2Cl_2}, {$MgSO_4$}
	min.		max.	
bl + ha	6.3	-	29.0	
ha + lw	29.0	-	53.8	
ha + ks	53.8	-	112.2	

	Input: {$MgCl_2$}			Output: {Na_2SO_4}, {$MgSO_4$}
	min.		max.	
bl + hx	0	-	25.7	

	Input: {$MgSO_4$}			Output: {Na_2Cl_2}, {Na_2SO_4}
	min.		max.	
bl + th	24.3	-	29.7	
th + vh	19.6	-	24.3	

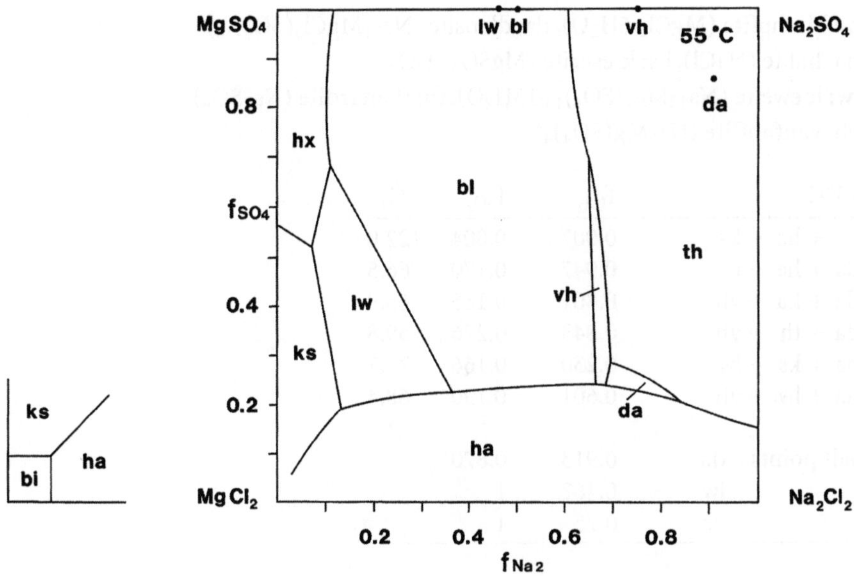
ks
ha
bi
Mg SO4
Na2SO4
Mg Cl2
Na2Cl2
lw
bl
vh
55 °C
da
0.8
hx
bl
fSO4
th
lw
vh
0.4
ks
0.2
da
ha
0.2
0.4
0.6
0.8
fNa2

III.4.4
Na_2Cl_2-$MgCl_2$-Na_2SO_4-$MgSO_4$-H_2O

bi: bischofite ($MgCl_2{\cdot}6H_2O$), da: d'ansite ($Na_{21}MgCl_3(SO_4)_{10}$)
ha: halite (NaCl), ks: kieserite ($MgSO_4{\cdot}H_2O$)
lw: loeweite ($Na_{12}Mg_7(SO_4)_{13}{\cdot}15H_2O$), th: thenardite ($Na_2SO_4$)
vh: vanthoffite ($Na_6Mg(SO_4)_4$)

75°C		f_{Na_2}	f_{SO_4}	C_T
bi + ha + ks		0.007	0.004	122.0
da + ha + th		0.947	0.170	66.5
da + ha + vh		0.801	0.165	65.9
da + th + vh		0.843	0.276	59.8
ha + ks + lw		0.230	0.166	75.7
ha + lw + vh		0.601	0.190	68.4
salt points:	da	0.913	0.870	
	lw	0.462	1	
	vh	0.75	1	

SALSYS

	Input: {$MgCl_2$}			Output: {Na_2Cl_2}, {Na_2SO_4}
	min.		max.	
da + ha	3.5	–	13.1	

	Input: {$MgCl_2$}			Output: {Na_2Cl_2}, {$MgSO_4$}
	min.		max.	
ha + ks	45.7	–	120.6	
ha + lw	14.3	–	45.7	
ha + vh	2.2	–	14.3	
ks + lw	–8.9	–	45.7	

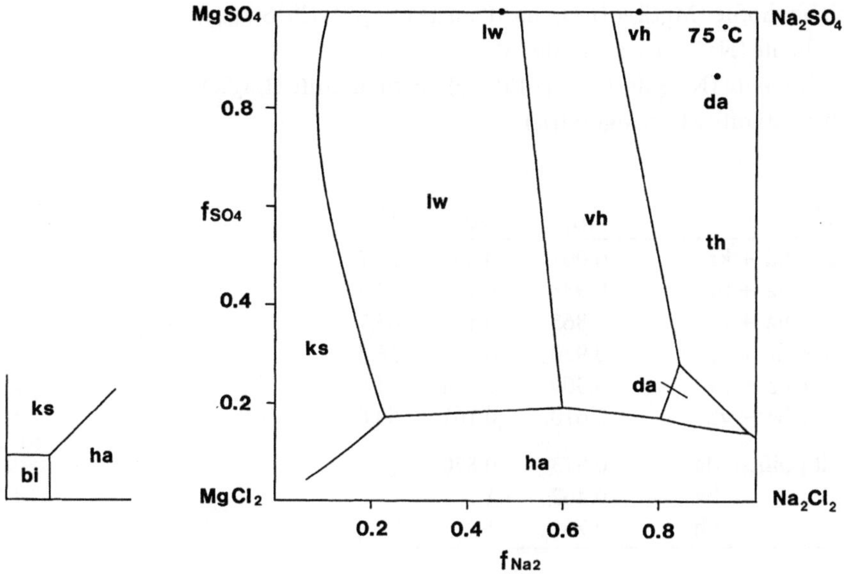

MgSO4
Na2SO4
75 °C
lw
vh
da
th
ks
ha
bi
0.8
0.4
0.2
fSO4
MgCl2
Na2Cl2
0.2
0.4
0.6
0.8
fNa2

III.4.4
Na_2Cl_2-$MgCl_2$-Na_2SO_4-$MgSO_4$-H_2O

bi: bischofite ($MgCl_2 \cdot 6H_2O$), da: d'ansite ($Na_{21}MgCl_3(SO_4)_{10}$)
ha: halite (NaCl), ks: kieserite ($MgSO_4 \cdot H_2O$)
lw: loeweite ($Na_{12}Mg_7(SO_4)_{13} \cdot 15H_2O$), th: thenardite ($Na_2SO_4$)
vh: vanthoffite ($Na_6Mg(SO_4)_4$)

90°C	f_{Na_2}	f_{SO_4}	C_T
bi + ha + ks	0.007	0.004	129.7
da + ha + th	0.988	0.137	65.8
da + ha + vh	0.865	0.135	65.7
da + th + vh	0.920	0.257	58.4
ha + ks + lw	0.309	0.154	74.5
ha + lw + vh	0.610	0.161	68.8
salt points: da	0.913	0.870	
lw	0.462	1	
vh	0.75	1	

SALSYS

	Input: {$MgCl_2$}			Output: {Na_2Cl_2}, {Na_2SO_4}
	min.		max.	
da + ha	0.8	–	8.9	
ha + lw	26.8	–	51.5	
ha + vh	8.9	–	26.8	

	Input: {Na_2Cl_2}			Output: {$MgCl_2$}, {Na_2SO_4}
	min.		max.	
ha + ks	0.4	–	11.5	

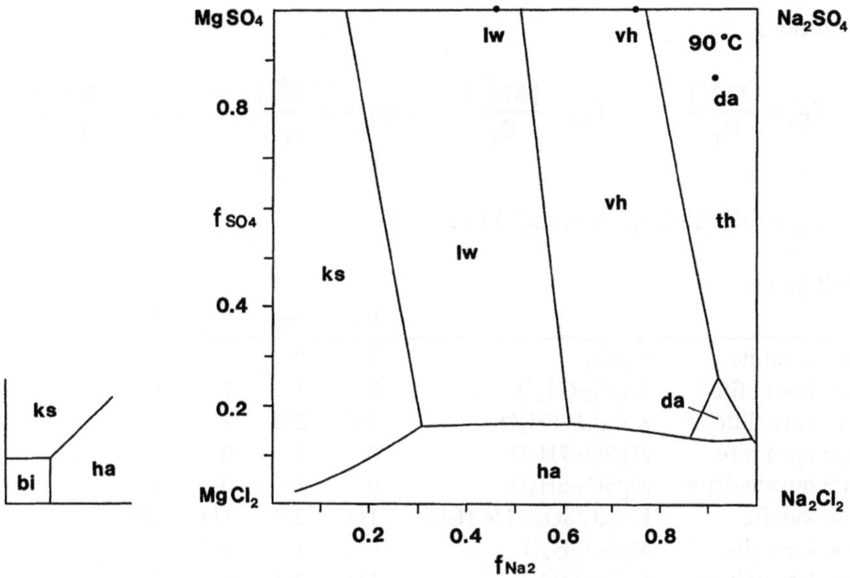

MgSO4
Na2SO4
MgCl2
Na2Cl2
90 °C
0.8
0.4
0.2
fSO4
0.2
0.4
0.6
0.8
fNa2
lw
vh
da
th
ks
ha
bi

III.4.5
K_2Cl_2-$MgCl_2$-K_2SO_4-$MgSO_4$-H_2O

Solutions

$$f_{K_2}=\frac{\{K_2^{2+}\}}{C_T} \qquad f_{Mg}=\frac{\{Mg^{2+}\}}{C_T} \qquad f_{Cl_2}=\frac{\{Cl_2^{2-}\}}{C_T} \qquad f_{SO_4}=\frac{\{SO_4^{2-}\}}{C_T}$$

$$C_T=\{K_2^{2+}\}+\{Mg^{2+}\}=\{Cl_2^{2-}\}+\{SO_4^{2-}\}$$

Salt points

		f_{K_2}	f_{Mg}	f_{Cl_2}	f_{SO_4}
ac: arcanite	K_2SO_4	1	0	0	1
bi: bischofite	$MgCl_2·6H_2O$	0	1	1	0
ca: carnallite	$KMgCl_3·6H_2O$	1/3	2/3	1	0
ep: epsomite	$MgSO_4·7H_2O$	0	1	0	1
hx: hexahydrite	$MgSO_4·6H_2O$	0	1	0	1
ka: kainite	$KMgClSO_4·11/4\ H_2O$	1/3	2/3	1/3	2/3
ks: kieserite	$MgSO_4·H_2O$	0	1	0	1
lg: langbeinite	$K_2Mg_2(SO_4)_3$	1/3	2/3	0	1
le: leonite	$K_2Mg(SO_4)_2·4H_2O$	1/2	1/2	0	1
pc: picromerite	$K_2Mg(SO_4)_2·6H_2O$	1/2	1/2	0	1
sy: sylvite	KCl	1	0	1	0

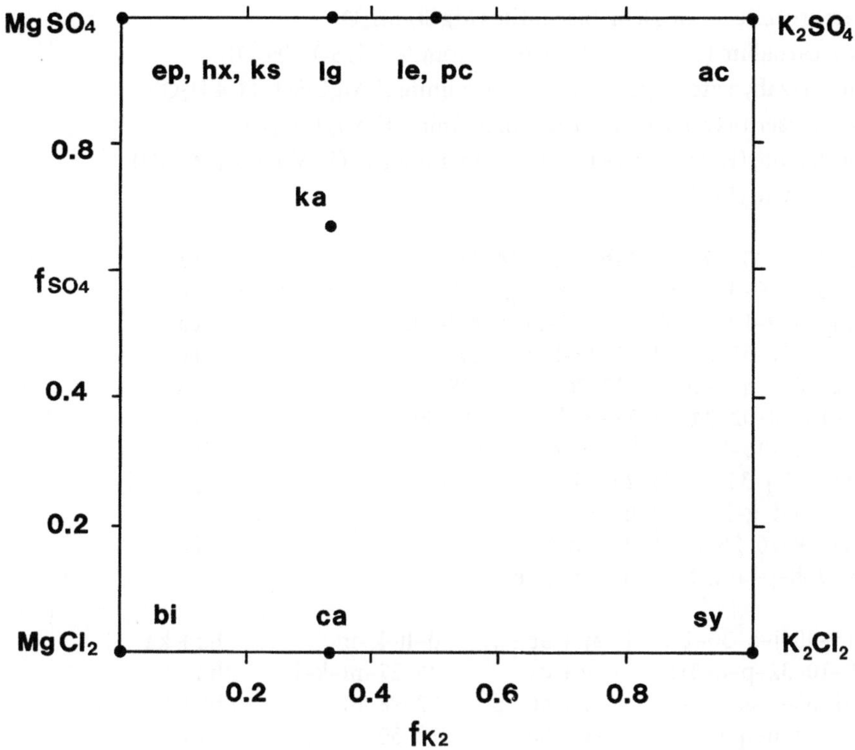

Mg SO4
K2SO4
ep, hx, ks
lg
le, pc
ac
0.8
ka
fSO4
0.4
0.2
bi
ca
sy
Mg Cl2
K2Cl2
0.2
0.4
0.6
0.8
fK2

III.4.5 K_2Cl_2-$MgCl_2$-K_2SO_4-$MgSO_4$-H_2O

ac: arcanite (K_2SO_4), bi: bischofite ($MgCl_2 \cdot 6H_2O$)
ca: carnallite ($KMgCl_3 \cdot 6H_2O$), ep: epsomite ($MgSO_4 \cdot 7H_2O$)
hx: hexahydrite ($MgSO_4 \cdot 6H_2O$), ka: kainite ($KMgClSO_4 \cdot 11/4\ H_2O$)
ks: kieserite ($MgSO_4 \cdot H_2O$), lg: langbeinite ($K_2Mg_2(SO_4)_3$)
le: leonite ($K_2Mg(SO_4)_2 \cdot 4H_2O$), pc: picromerite ($K_2Mg(SO_4)_2 \cdot 6H_2O$)
sy: sylvite (KCl)

1–2–12–30–29–11–37–b–c–38–14–13:	bi
5–6–10–9–31–a–p–32–34–q–n–g–e–33–35–j–o–36–16–15:	sy
14–13–9–10–32–38–c–b–37–31–a–p–h–d:	ca
3–11–37–31–33–19–25–26–i–d–b–29–21:	ep
22–30–c–h–k–m–28–27–26–i–d–b–29–21:	hx
4–12–38–32–34–20–28–m–k–h–c–30–22:	ks
7–8–16–15–35–j–o–36–18–24–23–17:	ac
33–e–f–j–35–17–23–25–19:	pc
23–25–f–g–j–o–n–l–m–27–24:	le
24–18–20–28–27–m–l–n–o–36–34–q:	lg
a–d–h–p–a–e–f–i–k–l–q–n–g–e:	ka

15–35–j–o–36–16:	sy + ac	d–h–k–i:	hx + ka
9–10–32–p–a–31:	sy + ca	26–27–m–k–i:	hx + le
31–33–e–a:	sy + ep	27–28–m:	hx + lg
a–e–g–n–q–p:	sy + ka	12–30–c–38:	ks + bi
32–34–q–p:	sy + ks	32–38–c–h–p:	ks + ca
33–35–j–g–e:	sy + pc	h–k–l–q–p:	ks + ka
g–j–o–n:	sy + le	k–l–m:	ks + le
34–36–o–n–q:	sy + lg	20–28–m–l–q–34:	ks + lg
11–29–b–37:	ep + bi	17–23–j–35:	ac + pc
31–37–b–d–a:	ep + ca	23–24–o–j:	ac + le
21–29–b–d–i–26:	ep + hx	18–24–o–36:	ac + lg
a–d–i–f–e:	ep + ka	e–f–g:	ka + pc
19–25–f–e–33:	ep + pc	f–g–n–l–k–i:	ka + le
25–26–i–f:	ep + le	l–q–n:	ka + lg
29–30–c–b:	hx + bi	24–27–m–l–n–o:	le + lg
b–c–h–d:	ca + hx	23–25–f–g–j:	le + pc
22–30–c–h–k–m–28:	ks + hx	13–14–38–37:	bi + ca

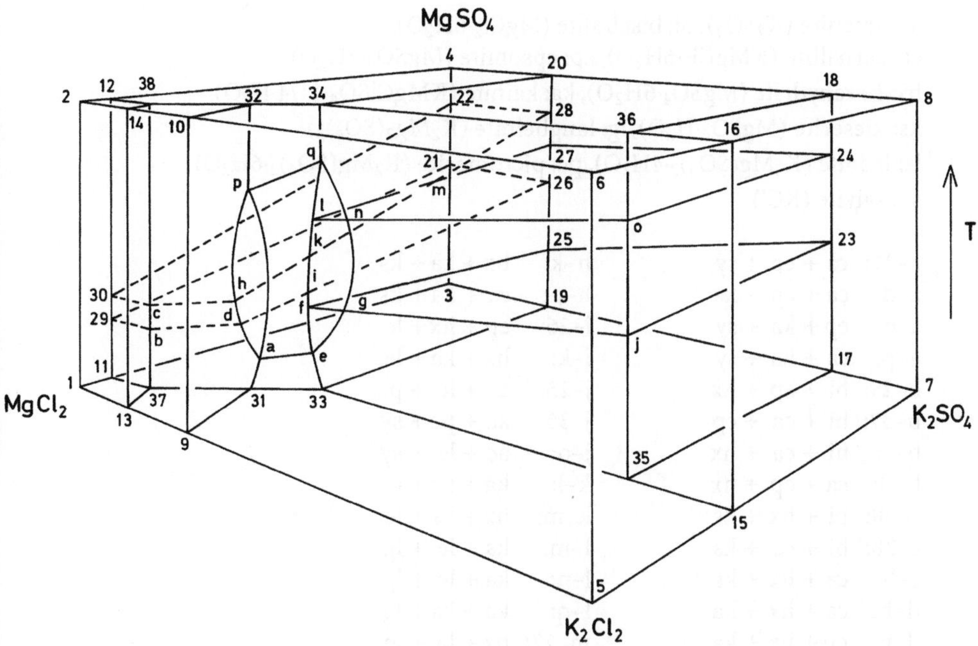
MgSO4
MgCl2
K2SO4
K2Cl2
T
1
2
3
4
5
6
7
8
9
10
11
12
13
14
15
16
17
18
19
20
21
22
23
24
25
26
27
28
29
30
31
32
33
34
35
36
37
38
a
b
c
d
e
f
g
h
i
j
k
l
m
n
o
p
q

III.4.5
K_2Cl_2-$MgCl_2$-K_2SO_4-$MgSO_4$-H_2O

ac: arcanite (K_2SO_4), bi: bischofite ($MgCl_2 \cdot 6H_2O$)
ca: carnallite ($KMgCl_3 \cdot 6H_2O$), ep: epsomite ($MgSO_4 \cdot 7H_2O$)
hx: hexahydrite ($MgSO_4 \cdot 6H_2O$), ka: kainite ($KMgClSO_4 \cdot 11/4\ H_2O$)
ks: kieserite ($MgSO_4 \cdot H_2O$), lg: langbeinite ($K_2Mg_2(SO_4)_3$)
le: leonite ($K_2Mg(SO_4)_2 \cdot 4H_2O$), pc: picromerite ($K_2Mg(SO_4)_2 \cdot 6H_2O$)
sy: sylvite (KCl)

a–31:	ca + ep + sy	h–k:	hx + ka + ks
a–d:	ca + ep + ka	h–p:	ca + ka + ks
a–e:	ep + ka + sy	i–26:	ep + hx + le
a–p:	ca + ka + sy	i–k:	hx + ka + le
b–29:	bi + ep + hx	j–23:	ac + le + pc
b–37:	bi + ca + ep	j–35:	ac + pc + sy
b–c:	bi + ca + hx	j–o:	ac + le + sy
b–d:	ca + ep + hx	k–l:	ka + ks + le
c–30:	bi + hx + ks	k–m:	hx + ks + le
c–38:	bi + ca + ks	l–m:	ks + le + lg
c–h:	ca + hx + ks	l–n:	ka + le + lg
d–h:	ca + hx + ka	l–q:	ka + ks + lg
d–i:	ep + hx + ka	m–27:	hx + le + lg
e–33:	ep + pc + sy	m–28:	hx + ks + lg
e–f:	ep + ka + pc	n–o:	le + lg + sy
e–g:	ka + pc + sy	n–q:	ka + lg + sy
f–25:	ep + le + pc	o–24:	ac + le + lg
f–g:	ka + le + pc	o–36:	ac + lg + sy
f–i:	ep + ka + le	p–32:	ca + ks + sy
g–n:	ka + le + sy	p–q:	ka + ks + sy
g–j:	le + pc + sy	q–34:	ks + lg + sy

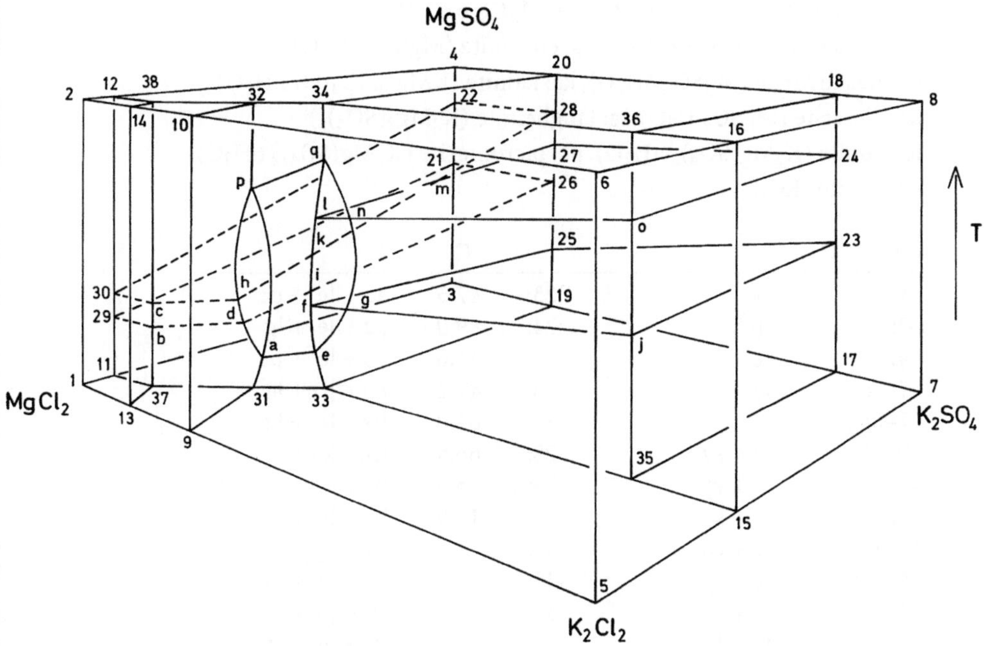

MgSO4
MgCl2
K2Cl2
K2SO4
T

III.4.5
K_2Cl_2-$MgCl_2$-K_2SO_4-$MgSO_4$-H_2O

ac: arcanite (K_2SO_4), bi: bischofite ($MgCl_2 \cdot 6H_2O$)
ca: carnallite ($KMgCl_3 \cdot 6H_2O$), ep: epsomite ($MgSO_4 \cdot 7H_2O$)
hx: hexahydrite ($MgSO_4 \cdot 6H_2O$), ka: kainite ($KMgClSO_4 \cdot 11/4\ H_2O$)
ks: kieserite ($MgSO_4 \cdot H_2O$), lg: langbeinite ($K_2Mg_2(SO_4)_3$)
le: leonite ($K_2Mg(SO_4)_2 \cdot 4H_2O$), pc: picromerite ($K_2Mg(SO_4)_2 \cdot 6H_2O$)
sy: sylvite (KCl)

	°C			°C	
1–2:		bi	23:	47.5	ac + le + pc
3:		ep	24:	89.0	ac + le + lg
4:		ks	25:	41.0	ep + le + pc
5–6:		sy	26:	47.2	ep + hx + le
7–8:		ac	27:	61.0	hx + le + lg
9–10:		ca + sy	28:	66.5	hx + ks + lg
11:		bi + ep	29:	13.0	bi + ep + hx
12:		bi + ks	30:	18.0	bi + hx + ks
13–14:		bi + ca	31:		ca + ep + sy
15–16:		ac + sy	32:		ca + ks + sy
17:		ac + pc	33:		ep + pc + sy
18:		ac + lg	34:		ks + lg + sy
19:		ep + pc	35:		ac + pc + sy
20:		ks + lg	36:		ac + lg + sy
21:	48.2	ep + hx	37:		bi + ca + ep
22:	67.5	hx + ks	38:		bi + ca + ks
a:	13	ca + ep + ka + sy	j:	35	ac + le + pc + sy
b:	13	bi + ca + ep + hx	k:	40	hx + ka + ks + le
c:	18	bi + ca + hx + ks	l:	52	ka + ks + le + lg
d:	18	ca + ep + hx + ka	m:	57	hx + ks + le + lg
e:	22.5	ep + ka + pc + sy	n:	63	ka + le + lg + sy
f:	23	ep + ka + le + pc	o:	65	ac + le + lg + sy
g:	25	ka + le + pc + sy	p:	76	ca + ka + ks + sy
h:	25	ca + hx + ka + ks	q:	85	ka + ks + lg + sy
i:	26	ep + hx + ka + le			

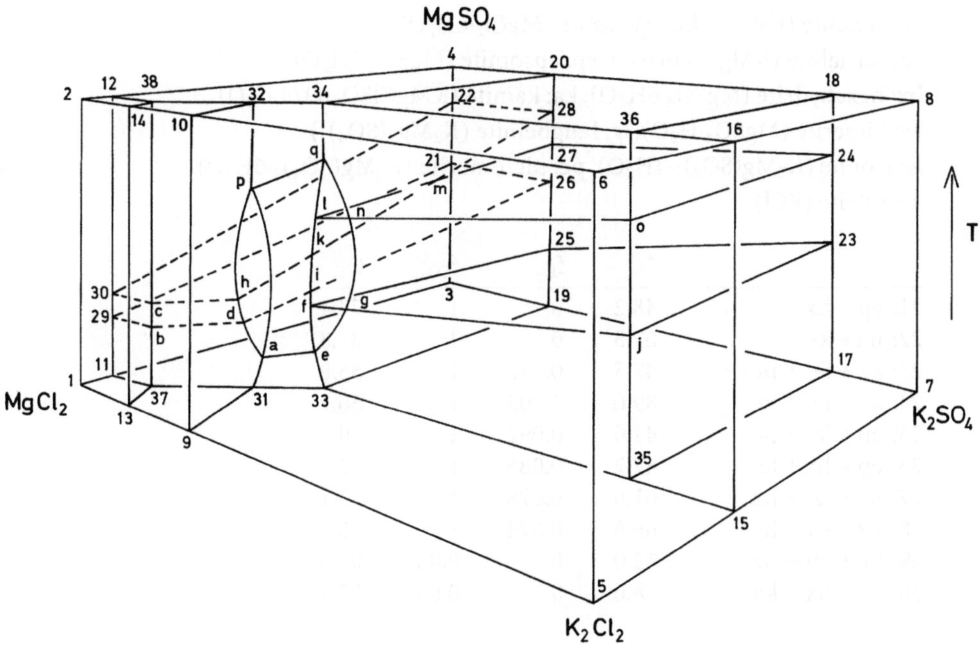

Mg SO4
4
20
18
2
12
38
32
34
22
28
8
14
10
36
16
q
21
27
24
p
m
26
6
l
n
o
k
25
23
T
h
i
30
3
19
29
c
d
f
g
b
a
e
j
17
11
1
7
MgCl2
37
31
33
K2SO4
13
9
35
15
5
K2Cl2

III.4.5
K_2Cl_2-$MgCl_2$-K_2SO_4-$MgSO_4$-H_2O

ac: arcanite (K_2SO_4), bi: bischofite ($MgCl_2 \cdot 6H_2O$)
ca: carnallite ($KMgCl_3 \cdot 6H_2O$), ep: epsomite ($MgSO_4 \cdot 7H_2O$)
hx: hexahydrite ($MgSO_4 \cdot 6H_2O$), ka: kainite ($KMgClSO_4 \cdot 11/4\ H_2O$)
ks: kieserite ($MgSO_4 \cdot H_2O$), lg: langbeinite ($K_2Mg_2(SO_4)_3$)
le: leonite ($K_2Mg(SO_4)_2 \cdot 4H_2O$), pc: picromerite ($K_2Mg(SO_4)_2 \cdot 6H_2O$)
sy: sylvite (KCl)

	°C	f_{K_2}	f_{SO_4}	C_T
21: ep + hx	48.2	0	1	73.4
22: hx + ks	67.5	0	1	87.8
23: ac + le + pc	47.5	0.342	1	55.0
24: ac + le + lg	89.0	0.405	1	66.7
25: ep + le + pc	41.0	0.092	1	79.3
26: ep + hx + le	47.2	0.086	1	82.4
27: le + lg + hx	61.0	0.078	1	93.1
28: hx + ks + lg	66.5	0.074	1	95.9
29: bi + ep + hx	13.0	0	0.057	105.8
30: bi + hx + ks	18.0	0	0.062	107.4
a: ca + ep + ka + sy	13	0.049	0.086	85.0
b: bi + ca + ep + hx	13	0.019	0.056	108
c: bi + ca + hx + ks	18	0.018	0.064	109
d: ca + ep + hx + ka	18	0.031	0.093	97
e: ep + ka + pc + sy	22.5	0.112	0.181	76.2
f: ep + ka + le + pc	23	0.104	0.210	74.2
g: ka + le + pc + sy	25	0.122	0.188	75.4
h: ca + hx + ka + ks	25	0.030	0.101	99.0
i: ep + hx + ka + le	26	0.049	0.185	81
j: ac + le + pc + sy	35	0.448	0.180	60.0
k: hx + ka + ks + le	40	0.051	0.253	79
l: ka + ks + le + lg	52	0.053	0.307	75
m: hx + ks + le + lg	57	0.107	0.640	75
n: ka + le + lg + sy	63	0.493	0.159	69
o: ac + le + lg + sy	65	0.529	0.162	68
p: ca + ka + ks + sy	76	0.101	0.030	99.0
q: ka + ks + lg + sy	85	0.124	0.036	98.4

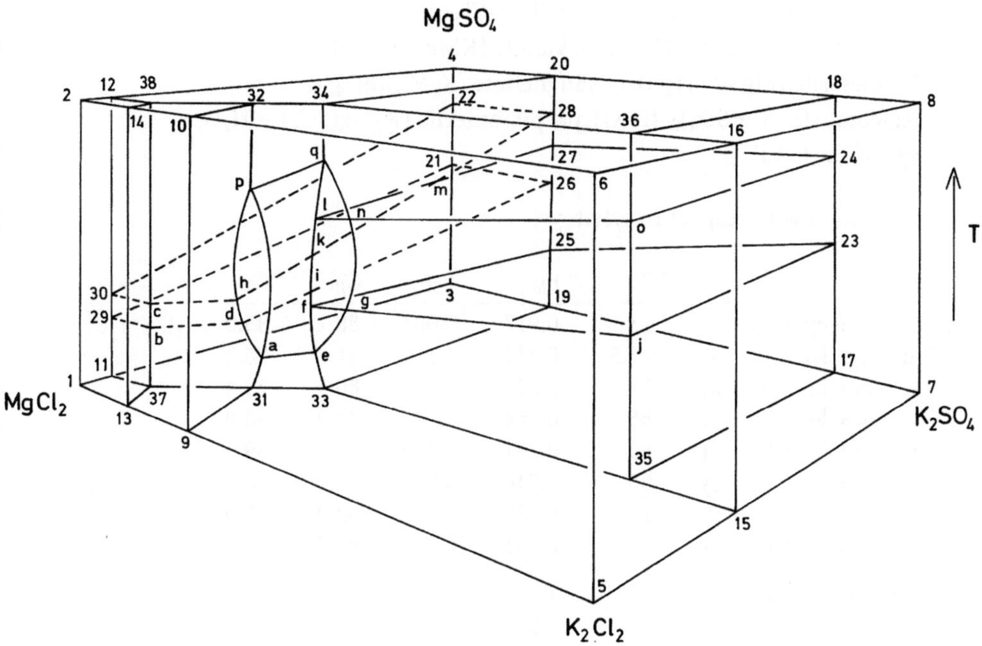

MgSO₄
MgCl₂
K₂SO₄
K₂Cl₂
T
1
2
3
4
5
6
7
8
9
10
11
12
13
14
15
16
17
18
19
20
21
22
23
24
25
26
27
28
29
30
31
32
33
34
35
36
37
38
a
b
c
d
e
f
g
h
i
j
k
l
m
n
o
p
q

III.4.5
K_2Cl_2-$MgCl_2$-K_2SO_4-$MgSO_4$-H_2O

ac: arcanite (K_2SO_4), ca: carnallite ($KMgCl_3 \cdot 6H_2O$)
ep: epsomite ($MgSO_4 \cdot 7H_2O$), ka: kainite ($KMgClSO_4 \cdot 11/4\ H_2O$)
ks: kieserite ($MgSO_4 \cdot H_2O$), lg: langbeinite ($K_2Mg_2(SO_4)_3$)
le: leonite ($K_2Mg(SO_4)_2 \cdot 4H_2O$), pc: picromerite ($K_2Mg(SO_4)_2 \cdot 6H_2O$)
sy: sylvite (KCl)

Saturation with respect to sylvite

		°C	F_{Mg}	F_{SO_4}	C_P	C_T
a:	ca + ep + ka + sy	13	0.917	0.083	88.1	85.0
e:	ep + ka + pc + sy	22.5	0.831	0.169	81.5	76.2
g:	ka + le + pc + sy	25	0.823	0.177	80.4	75.4
j:	ac + le + pc + sy	35	0.754	0.246	43.9	60.0
n:	ka + le + lg + sy	63	0.761	0.239	46	69
o:	ac + le + lg + sy	65	0.744	0.256	43	68
p:	ca + k + ks + sy	76	0.967	0.033	92.0	99.0
q:	ka + ks + lg + sy	85	0.961	0.039	89.7	98.4
salt points:	ca		1	0		
	ka		0.5	0.5		
	lg		0.4	0.6		
	le, pc		0.333	0.667		

$$F_{Mg} = \frac{\{Mg^{2+}\}}{C_P} \qquad F_{SO_4} = \frac{\{SO_4^{2-}\}}{C_P}$$

$$C_P = \{Mg^{2+}\} + \{SO_4^{2-}\}$$

$$C_T = \{K_2^{2+}\} + \{Mg^{2+}\} = \{Cl_2^{2-}\} + \{SO_4^{2-}\}$$

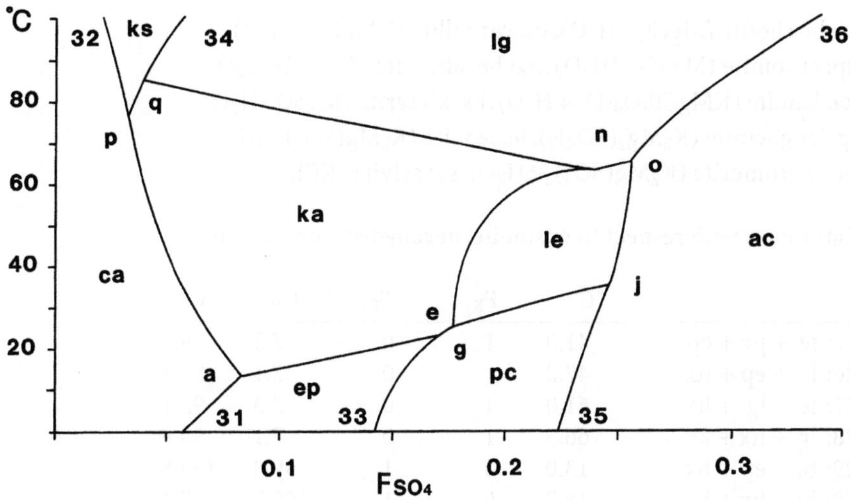

°C
80
60
40
20
32
ks
34
lg
36
q
p
n
o
ka
le
ac
ca
j
e
g
a
pc
ep
31
33
35
0.1
0.2
0.3
FSO4

III.4.5
K_2Cl_2-$MgCl_2$-K_2SO_4-$MgSO_4$-H_2O

bi: bischofite ($MgCl_2 \cdot 6H_2O$), ca: carnallite ($KMgCl_3 \cdot 6H_2O$)
ep: epsomite ($MgSO_4 \cdot 7H_2O$), hx: hexahydrite ($MgSO_4 \cdot 6H_2O$)
ka: kainite ($KMgClSO_4 \cdot 11/4\ H_2O$), ks: kieserite ($MgSO_4 \cdot H_2O$)
lg: langbeinite ($K_2Mg_2(SO_4)_3$), le: leonite ($K_2Mg(SO_4)_2 \cdot 4H_2O$)
pc: picromerite ($K_2Mg(SO_4)_2 \cdot 6H_2O$), sy: sylvite (KCl)

Saturation with respect to epsomite, hexahydrite, or kieserite

	°C	F_{K_2}	F_{Cl_2}	C_P	C_T
25: le + pc + ep	41.0	1	0	7.3	79.25
26: le + ep + hx	47.2	1	0	7.1	82.4
27: le + lg + hx	61.0	1	0	7.3	93.1
28: lg + hx + ks	66.5	1	0	7.1	95.9
29: bi + ep + hx	13.0	0	1	99.8	105.8
30: bi + hx + ks	18.0	0	1	100.7	107.4
a: ca + ka + sy + ep	13	0.051	0.949	81.9	85.0
b: bi + ca + ep + hx	13	0.019	0.981	104	108
c: bi + ca + hx + ks	18	0.019	0.981	104	109
d: ca + ka + ep + hx	18	0.033	0.967	91	97
e: ka + pc + sy + ep	22.5	0.120	0.880	70.9	76.2
f: ka + le + pc + ep	23	0.116	0.884	66.3	74.2
h: ca + ka + hx + ks	25	0.033	0.967	92.0	99.0
i: ka + le + ep + hx	26	0.057	0.943	70	81
k: ka + le + hx + ks	40	0.063	0.937	63	79
l: ka + le + lg + ks	52	0.071	0.929	56	75
m: le + lg + hx + ks	57	0.229	0.771	35	75
p: ca + ka + sy + ks	76	0.094	0.906	106.0	99.0
q: ka + lg + sy + ks	85	0.114	0.886	107.1	98.4
salt points: ca		0.25	0.75		
ka		0.5	0.5		
lg, le, pc		1	0		

$$F_{K_2} = \frac{\{K_2^{2+}\}}{C_P} \qquad F_{Cl_2} = \frac{\{Cl_2^{2-}\}}{C_P}$$

$$C_P = \{K_2^{2+}\} + \{Cl_2^{2-}\}$$

$$C_T = \{K_2^{2+}\} + \{Mg^{2+}\} = \{Cl_2^{2-}\} + \{SO_4^{2-}\}$$

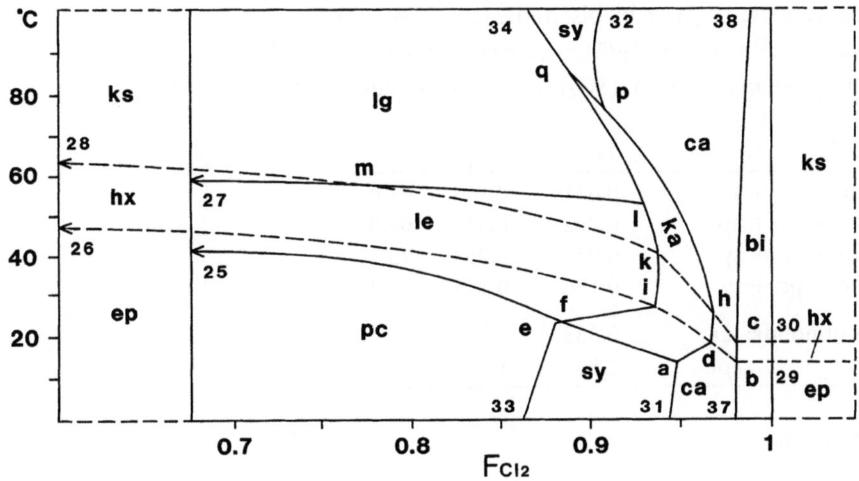

°C
80
60
40
20
ks
hx
ep
28
26
27
25
lg
m
le
pc
34
sy
32
38
q
p
ca
ks
l
ka
bi
k
i
h
f
e
c
30
hx
d
a
sy
b
29
ep
ca
33
31
37
0.7
0.8
0.9
1
F_{Cl_2}

III.4.5
K_2Cl_2-$MgCl_2$-K_2SO_4-$MgSO_4$-H_2O

ac: arcanite (K_2SO_4), bi: bischofite ($MgCl_2 \cdot 6H_2O$)
ca: carnallite ($KMgCl_3 \cdot 6H_2O$), ep: epsomite ($MgSO_4 \cdot 7H_2O$)
pc: picromerite ($K_2Mg(SO_4)_2 \cdot 6H_2O$), sy: sylvite (KCl)

0°C		f_{K_2}	f_{SO_4}	C_T
ac + pc + sy		0.431	0.164	47.6
bi + ca + ep		0.002	0.035	102.0
ca + ep + sy		0.056	0.058	75.5
ep + pc + sy		0.136	0.143	59.4
salt points:	ca	0.333	0	
	pc	0.5	1	

SALSYS

	Input: {$MgCl_2$}			Output: {K_2Cl_2}, {$MgSO_4$}
	min.		max.	
ac + pc	−9.6	–	19.3	
ac + sy	−1.3	–	19.3	
ep + pc	−3.5	–	42.8	
pc + sy	19.3	–	42.8	

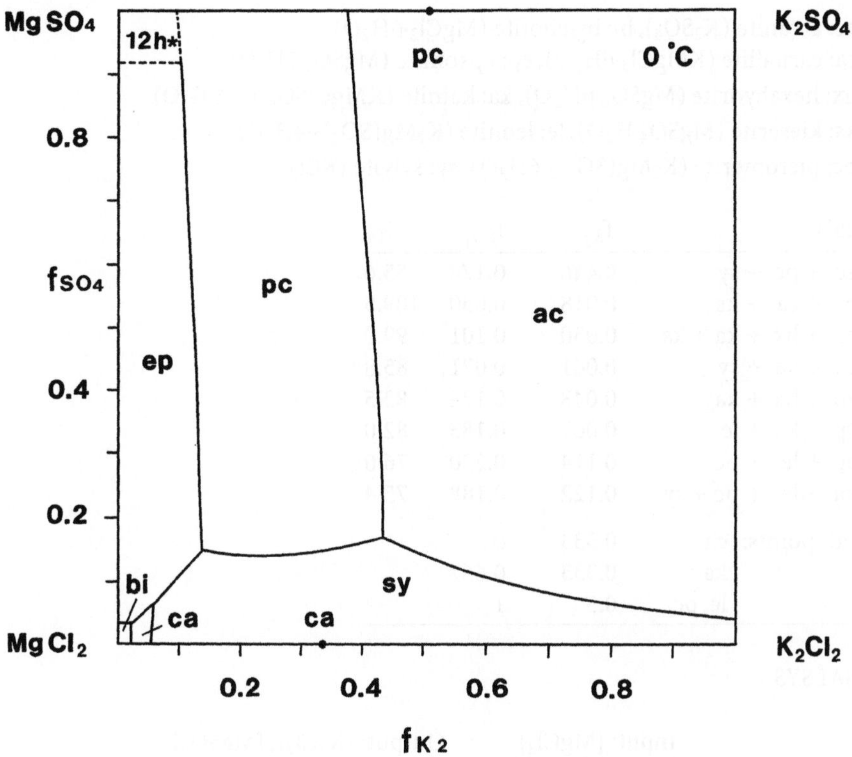
MgSO4
K2SO4
MgCl2
K2Cl2
0 °C
12h*
pc
pc
ac
ep
bi
ca
ca
sy
0.8
fSO4
0.4
0.2
0.2
0.4
0.6
0.8
fK2

III.4.5
K_2Cl_2-$MgCl_2$-K_2SO_4-$MgSO_4$-H_2O

ac: arcanite (K_2SO_4), bi: bischofite ($MgCl_2 \cdot 6H_2O$)
ca: carnallite ($KMgCl_3 \cdot 6H_2O$), ep: epsomite ($MgSO_4 \cdot 7H_2O$)
hx: hexahydrite ($MgSO_4 \cdot 6H_2O$), ka: kainite ($KMgClSO_4 \cdot 11/4\ H_2O$)
ks: kieserite ($MgSO_4 \cdot H_2O$), le: leonite ($K_2Mg(SO_4)_2 \cdot 4H_2O$)
pc: picromerite ($K_2Mg(SO_4)_2 \cdot 6H_2O$), sy: sylvite (KCl)

25°C		f_{K_2}	f_{SO_4}	C_T
ac + pc + sy		0.446	0.174	55.6
bi + ca + ks		0.018	0.050	109.5
ca + hx + ka + ks		0.030	0.101	99.0
ca + ka + sy		0.041	0.071	85.8
ep + hx + ka		0.048	0.174	83.5
ep + ka + le		0.061	0.183	82.0
ep + le + pc		0.114	0.220	76.0
ka + le + pc + sy		0.122	0.188	75.4
salt points:	ca	0.333	0	
	ka	0.333	0.667	
	le, pc	0.5	1	

SALSYS

	Input: {$MgCl_2$}			Output: {K_2Cl_2}, {$MgSO_4$}
	min.		max.	
ac + pc	−14.7	–	21.1	
ac + sy	−1.5	–	21.1	
ep + pc	−6.1	–	50.6	
ka + sy	52.0	–	76.2	
pc + sy	21.1	–	52.0	

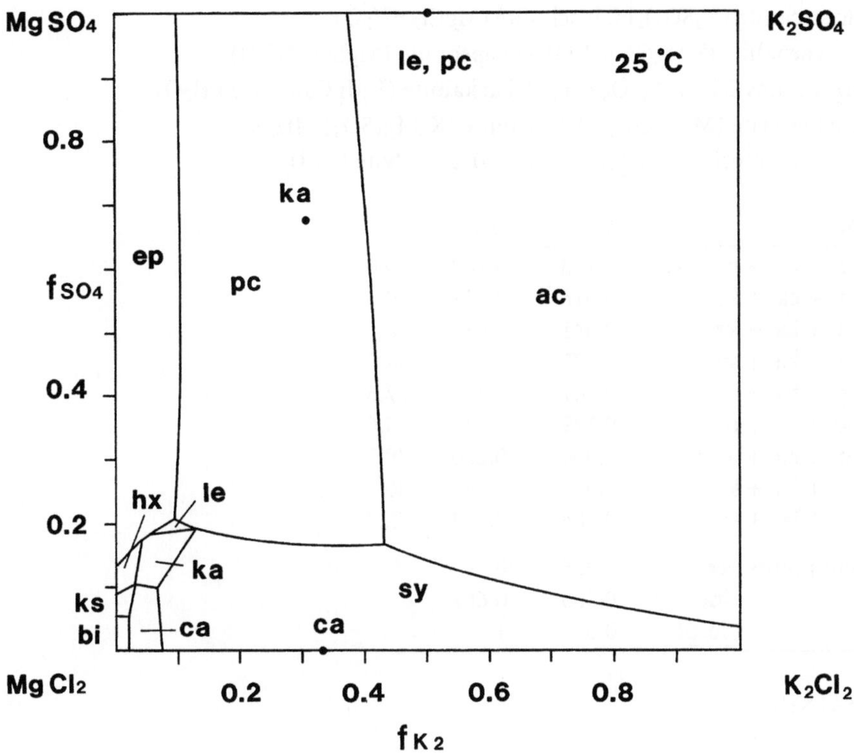

Mg SO4
K2SO4
le, pc
25 °C
0.8
ka
ep
fSO4
pc
ac
0.4
hx
le
0.2
ka
sy
ks
bi
ca
ca
Mg Cl2
0.2
0.4
0.6
0.8
K2Cl2
fK2

III.4.5
K_2Cl_2-$MgCl_2$-K_2SO_4-$MgSO_4$-H_2O

ac: arcanite (K_2SO_4), bi: bischofite ($MgCl_2 \cdot 6H_2O$)
ca: carnallite ($KMgCl_3 \cdot 6H_2O$), ep: epsomite ($MgSO_4 \cdot 7H_2O$)
hx: hexahydrite ($MgSO_4 \cdot 6H_2O$), ka: kainite ($KMgClSO_4 \cdot 11/4\ H_2O$)
ks: kieserite ($MgSO_4 \cdot H_2O$), le: leonite ($K_2Mg(SO_4)_2 \cdot 4H_2O$)
pc: picromerite ($K_2Mg(SO_4)_2 \cdot 6H_2O$), sy: sylvite (KCl)

35°C		f_{K_2}	f_{SO_4}	C_T
ac + le + pc + sy		0.448	0.180	60.0
bi + ca + ks		0.018	0.027	110
ca + ka + ks		0.041	0.082	98
ca + ka + sy		0.077	0.056	84.2
ep + hx + le		0.101	0.345	77.3
ep + le + pc		0.129	0.473	74.6
hx + ka + le		0.054	0.270	79.7
hx + ka + ks		0.047	0.198	86
ka + le + sy		0.184	0.174	72.3
salt points:	ca	0.333	0	
	ka	0.333	0.667	
	le, pc	0.5	1	

SALSYS

	Input: {$MgCl_2$}			Output: {K_2Cl_2}, {$MgSO_4$}
	min.		max.	
ac + pc	−16.5	-	22.3	
ac + sy	−1.7	-	22.3	
ep + le	29.7	-	42.8	
ep + pc	−6.9	-	29.7	
ka + sy	46.4	-	73.0	
le + sy	22.3	-	46.4	

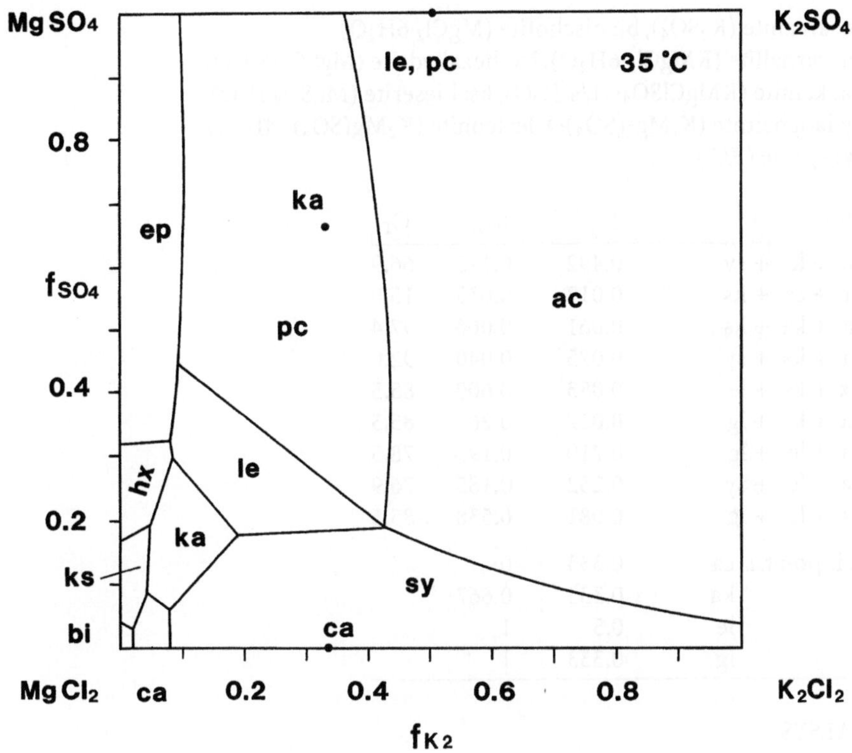
MgSO4
K2SO4
le, pc
35 °C
0.8
ka
ep
fSO4
ac
pc
0.4
le
hx
0.2
ka
sy
ks
ca
bi
MgCl2
ca
0.2
0.4
0.6
0.8
K2Cl2
fK2

III.4.5
K_2Cl_2-$MgCl_2$-K_2SO_4-$MgSO_4$-H_2O

ac: arcanite (K_2SO_4), bi: bischofite ($MgCl_2{\cdot}6H_2O$)
ca: carnallite ($KMgCl_3{\cdot}6H_2O$), hx: hexahydrite ($MgSO_4{\cdot}6H_2O$)
ka: kainite ($KMgClSO_4{\cdot}11/4\ H_2O$), ks: kieserite ($MgSO_4{\cdot}H_2O$)
lg: langbeinite ($K_2Mg_2(SO_4)_3$), le: leonite ($K_2Mg(SO_4)_2{\cdot}4H_2O$)
sy: sylvite (KCl)

55°C		f_{K_2}	f_{SO_4}	C_T
ac + le + sy		0.492	0.173	66.9
bi + ca + ks		0.017	0.033	115.0
ca + ka + ks		0.061	0.066	97.4
ca + ka + sy		0.075	0.040	92.0
hx + ks + le		0.083	0.600	85.5
ka + ks + lg		0.057	0.262	85.5
ka + le + lg		0.210	0.195	78.5
ka + le + sy		0.252	0.185	76.9
ks + le + lg		0.081	0.538	85.5
salt points:	ca	0.333	0	
	ka	0.333	0.667	
	le	0.5	1	
	lg	0.333	1	

SALSYS

	Input: {$MgCl_2$}			Output: {K_2Cl_2}, {$MgSO_4$}
	min.		max.	
ac + le	−19.8	–	22.4	
ac + sy	−1.9	–	22.4	
hx + le	−7.2	–	27.1	
ka + sy	43.3	–	85.1	
le + sy	22.4	–	43.3	

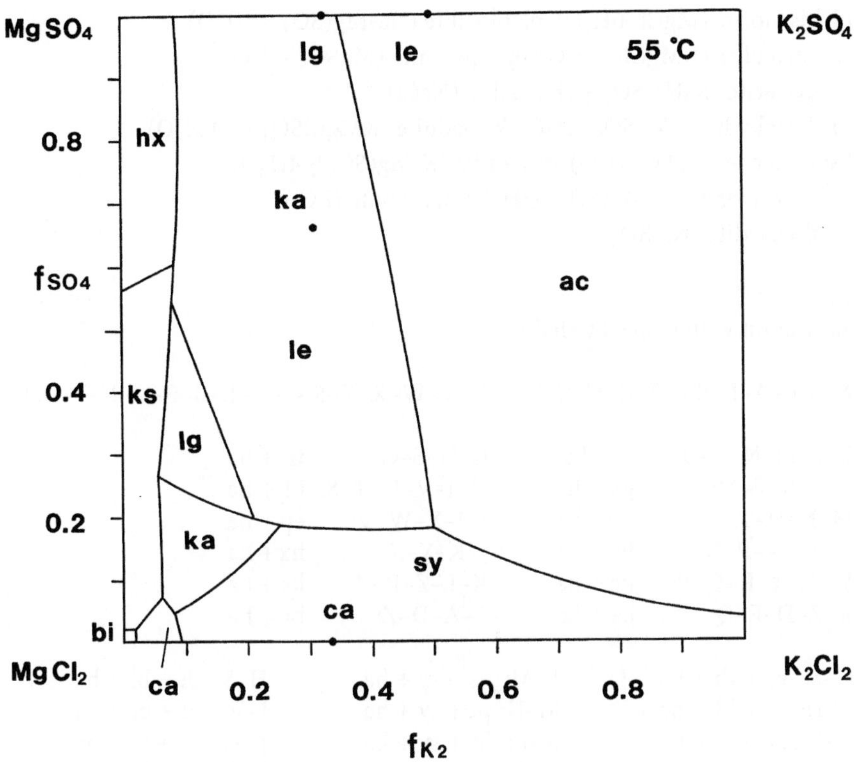
MgSO4
K2SO4
55 °C
lg
le
hx
ka
ac
le
ks
lg
ka
sy
ca
bi
MgCl2
K2Cl2
ca
0.8
fSO4
0.4
0.2
0.2
0.4
0.6
0.8
fK2

III.5
Na_2Cl_2-K_2Cl_2-$MgCl_2$-Na_2SO_4-K_2SO_4-$MgSO_4$-H_2O

bi: bischofite ($MgCl_2 \cdot 6H_2O$), bl: bloedite ($Na_2Mg(SO_4)_2 \cdot 4H_2O$)
ca: carnallite ($KMgCl_3 \cdot 6H_2O$), ep: epsomite ($MgSO_4 \cdot 7H_2O$)
gs: glaserite ($K_3Na(SO_4)_2$), ha: halite (NaCl)
hx: hexahydrite ($MgSO_4 \cdot 6H_2O$), ka: kainite ($KMgClSO_4 \cdot 11/4\ H_2O$)
ks: kieserite ($MgSO_4 \cdot H_2O$), le: leonite ($K_2Mg(SO_4)_2 \cdot 4H_2O$)
pc: picromerite ($K_2Mg(SO_4)_2 \cdot 6H_2O$), sy: sylvite (KCl)
th: thenardite (Na_2SO_4)

25°C
Saturation with respect to halite

Na_2Cl_2–A–L–K–J–I–H–C–G–S–T–U–V–W–X–Y–R–Z–D–E–B–F–M–N–P–Q: ha

B–F–M–N–P–Q–E:	sy + ha	C–H–S–G:	th + ha
F–G–S–T–M:	gs + ha	H–I–V–U–T–S:	bl + ha
M–N–U–T:	pc + ha	I–J–X–W–V:	ep + ha
U–V–W–P–N:	le + ha	J–K–Y–X:	hx + ha
W–X–Y–R–Q–P:	ka + ha	K–L–Z–R–Y:	ks + ha
R–Z–D–E–Q:	ca + ha	L–A–D–Z:	bi + ha

G–S:	gs + th + ha	F–M:	gs + sy + ha	H–S:	th + bl + ha
S–T:	gs + bl + ha	M–N:	pc + sy + ha	I–V:	bl + ep + ha
T–U:	pc + bl + ha	N–P:	le + sy + ha	J–X:	ep + hx + ha
U–V:	le + bl + ha	P–Q:	ka + sy + ha	K–Y:	hx + ks + ha
V–W:	le + ep + ha	Q–R:	ka + ca + ha	L–Z:	bi + ks + ha
W–X:	ka + ep + ha	R–Z:	ks + ca + ha	M–T:	gs + pc + ha
X–Y:	ka + hx + ha	N–U:	pc + le + ha		
Y–R:	ka + ks + ha	P–W:	le + ka + ha		

M:	gs + pc + sy + ha	S:	bl + gs + th + ha	W:	ep + ka + le + ha
N:	le + pc + sy + ha	T:	bl + gs + pc + ha	X:	ep + hx + ka + ha
P:	ka + le + sy + ha	U:	bl + le + pc + ha	Y:	hx + ka + ks + ha
Q:	ca + ka + sy + ha	V:	bl + ep + le + ha	Z:	bi + ca + ks + ha
R:	ca + ka + ks + ha				

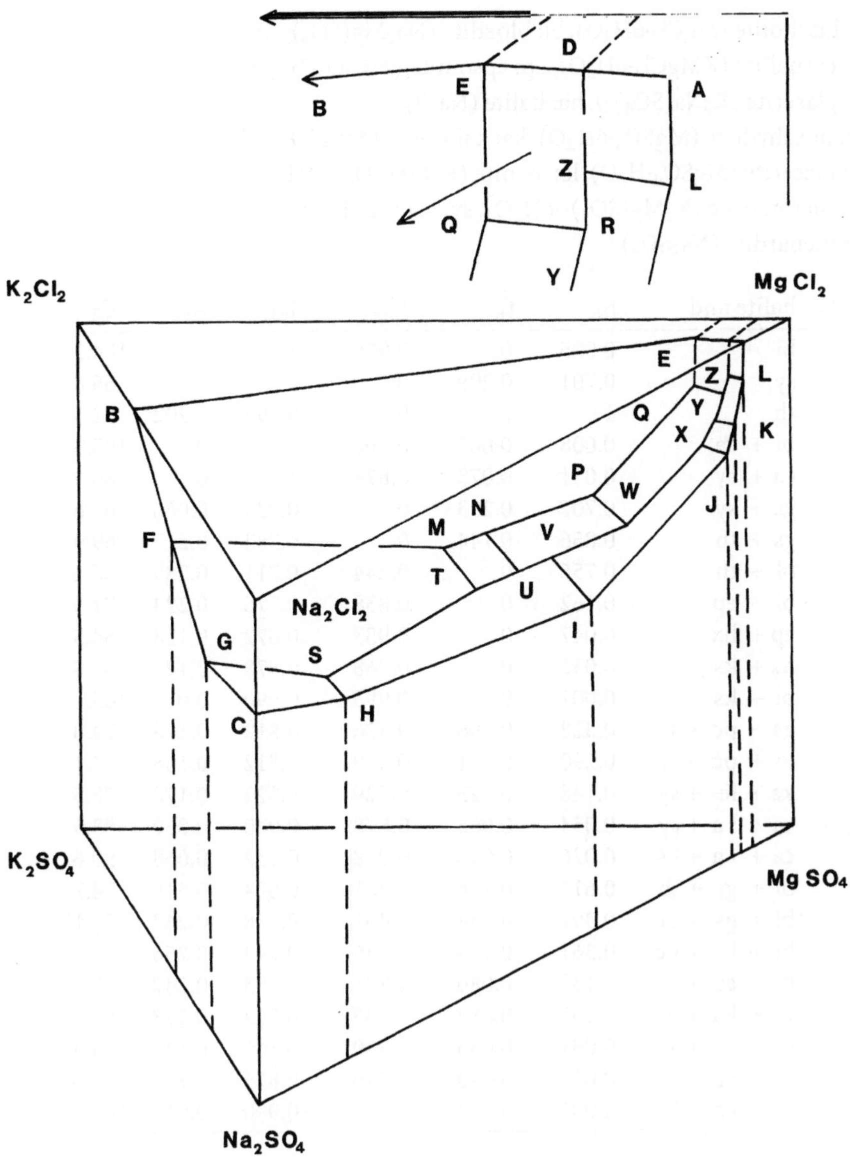
D
E
A
B
Z
L
Q
R
Y
K_2Cl_2
Mg Cl_2
E
Z
L
B
Q
Y
X
K
P
W
J
N
M
V
F
T
U
Na_2Cl_2
I
G
S
H
C
K_2SO_4
Mg SO_4
Na_2SO_4

III.5
Na_2Cl_2-K_2Cl_2-$MgCl_2$-Na_2SO_4-K_2SO_4-$MgSO_4$-H_2O

bi: bischofite ($MgCl_2{\cdot}6H_2O$), bl: bloedite ($Na_2Mg(SO_4)_2{\cdot}4H_2O$)
ca: carnallite ($KMgCl_3{\cdot}6H_2O$), ep: epsomite ($MgSO_4{\cdot}7H_2O$)
gs: glaserite ($K_3Na(SO_4)_2$), ha: halite (NaCl)
hx: hexahydrite ($MgSO_4{\cdot}6H_2O$), ka: kainite ($KMgClSO_4{\cdot}11/4\ H_2O$)
ks: kieserite ($MgSO_4{\cdot}H_2O$), le: leonite ($K_2Mg(SO_4)_2{\cdot}4H_2O$)
pc: picromerite ($K_2Mg(SO_4)_2{\cdot}6H_2O$), sy: sylvite (KCl)
th: thenardite (Na_2SO_4)

25°C	halite and	f_{Na_2}	f_{K_2}	f_{Mg}	f_{Cl_2}	f_{SO_4}	C_T
A	bi	0.005	0	0.995	1	0	104.2
B	sy	0.701	0.299	0	1	0	65.8
C	th	1	0	0	0.798	0.202	62.5
D	bi + ca	0.008	0.002	0.990	1	0	105.3
E	ca + sy	0.051	0.073	0.876	1	0	80.6
F	gs + sy	0.707	0.293	0	0.934	0.066	68.6
G	gs + th	0.856	0.144	0	0.783	0.217	69.0
H	bl + th	0.756	0	0.244	0.711	0.289	66.4
I	bl + ep	0.162	0	0.838	0.776	0.224	72.6
J	ep + hx	0.047	0	0.953	0.872	0.128	84.5
K	hx + ks	0.032	0	0.968	0.878	0.122	85.8
L	bi + ks	0.007	0	0.993	0.990	0.010	105.1
M	gs + pc + sy	0.328	0.196	0.476	0.812	0.188	74.0
N	le + pc + sy	0.290	0.181	0.529	0.812	0.188	75.1
P	ka + le + sy	0.148	0.123	0.729	0.822	0.178	78.5
Q	ca + ka + sy	0.054	0.068	0.878	0.940	0.060	83.3
R	ca + ka + ks	0.026	0.016	0.958	0.932	0.068	90.6
S	bl + gs + th	0.613	0.126	0.261	0.684	0.316	74.7
T	bl + gs + pc	0.391	0.158	0.451	0.738	0.262	74.1
U	bl + le + pc	0.361	0.153	0.486	0.744	0.256	74.6
V	bl + ep + le	0.139	0.080	0.781	0.753	0.247	79.8
W	ep + ka + le	0.131	0.084	0.785	0.772	0.228	81.0
X	ep + hx + ka	0.084	0.036	0.880	0.865	0.135	83.8
Y	hx + ka + ks	0.036	0.018	0.946	0.875	0.125	87.9
Z	bi + ca + ks	0.007	0.002	0.991	0.990	0.010	106.2

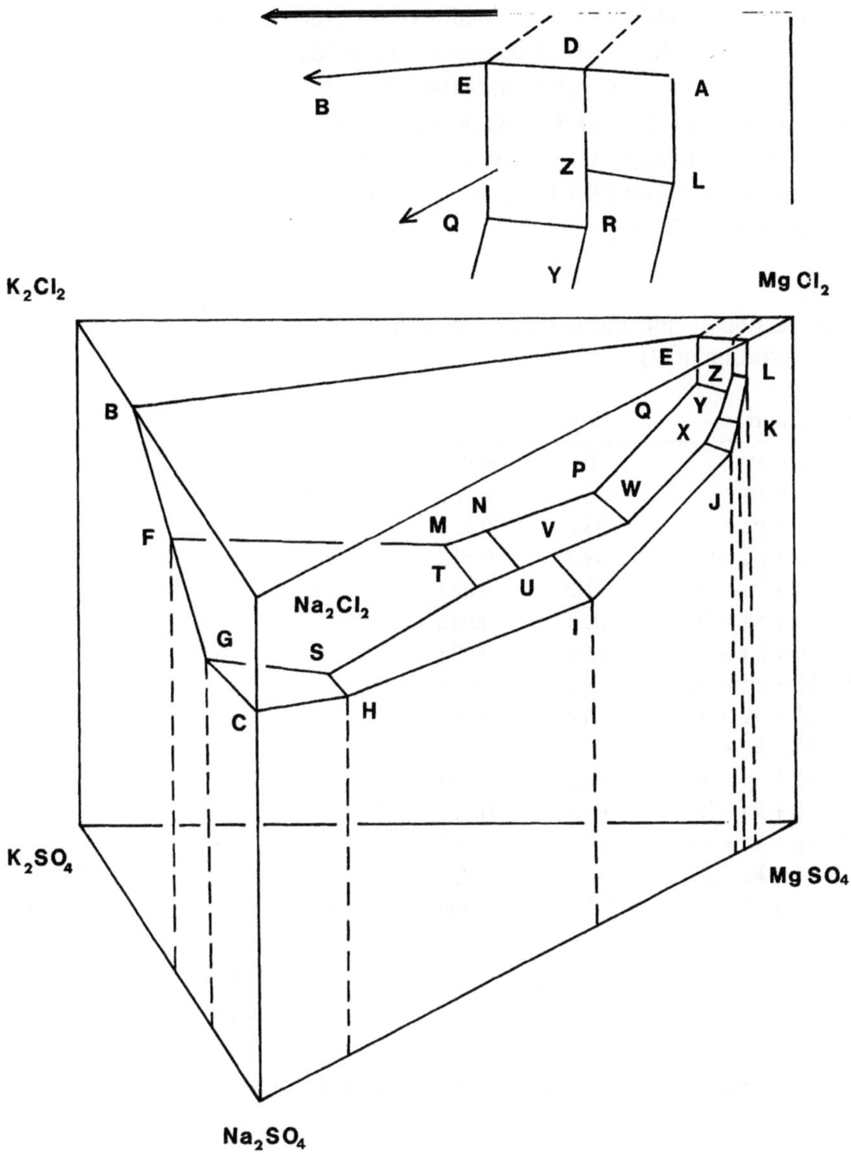
D
E
A
B
Z
L
Q
R
Y
K_2Cl_2
$MgCl_2$
E
Z
L
B
Q
Y
X
K
P
W
M
N
J
F
V
T
U
Na_2Cl_2
I
G
S
H
C
K_2SO_4
$MgSO_4$
Na_2SO_4

III.5
Na_2Cl_2-K_2Cl_2-$MgCl_2$-Na_2SO_4-K_2SO_4-$MgSO_4$-H_2O

bi: bischofite ($MgCl_2 \cdot 6H_2O$), bl: bloedite ($Na_2Mg(SO_4)_2 \cdot 4H_2O$)
ca: carnallite ($KMgCl_3 \cdot 6H_2O$), ep: epsomite ($MgSO_4 \cdot 7H_2O$)
gs: glaserite ($K_3Na(SO_4)_2$), ha: halite (NaCl)
hx: hexahydrite ($MgSO_4 \cdot 6H_2O$), ka: kainite ($KMgClSO_4 \cdot 11/4\ H_2O$)
ks: kieserite ($MgSO_4 \cdot H_2O$), le: leonite ($K_2Mg(SO_4)_2 \cdot 4H_2O$)
pc: picromerite ($K_2Mg(SO_4)_2 \cdot 6H_2O$), sy: sylvite (KCl)
th: thenardite (Na_2SO_4)

25°C
The fraction of chloride as a function of the total concentration of Mg^{2+}

halite and	$\{Mg^{2+}\}$	f_{Cl_2}
F: gs + sy	0	0.934
G: gs + th	0	0.783
M: gs + pc + sy	35.2	0.812
N: le + pc + sy	39.7	0.812
P: ka + le + sy	57.2	0.822
Q: ca + ka + sy	73.1	0.940
R: ca + ka + ks	86.8	0.932
S: bl + gs + th	19.5	0.684
T: bl + gs + pc	33.4	0.738
U: bl + le + pc	36.3	0.744
V: bl + ep + le	62.3	0.753
W: ep + ka + le	63.6	0.772
X: ep + hx + ka	73.8	0.865
Y: hx + ka + ks	83.1	0.875
Z: bi + ca + ks	105.3	0.990

$$\{Mg^{2+}\} = \{MgCl_2\} + \{MgSO_4\} \qquad f_{Cl_2} = \frac{\{Cl_2^{2-}\}}{C_T}$$

$$C_T = \{Na_2^{2+}\} + \{K_2^{2+}\} + \{Mg^{2+}\} = \{Cl_2^{2-}\} + \{SO_4^{2-}\}$$

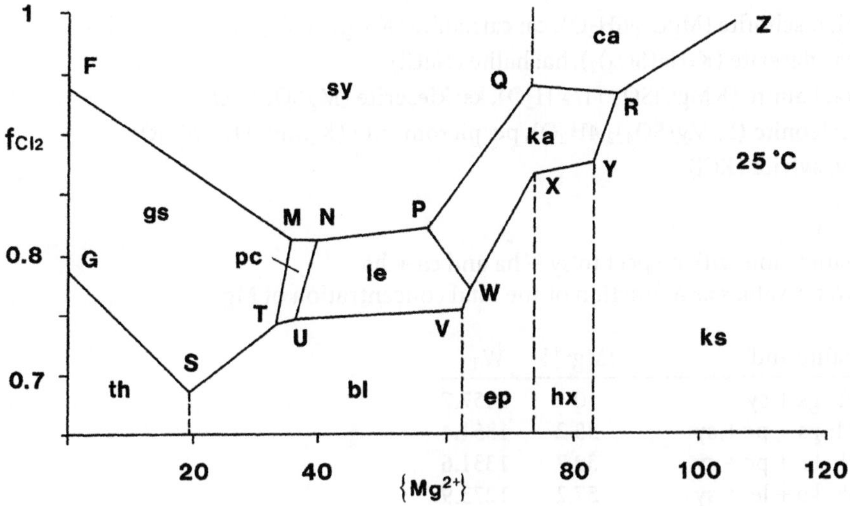

1
0.8
0.7
f_{Cl_2}
20
40
80
100
120
{Mg^{2+}}
F
G
S
T
U
M
N
P
Q
R
Z
W
V
X
Y
sy
gs
pc
le
th
bl
ep
hx
ka
ca
ks
25 °C

III.5
Na_2Cl_2-K_2Cl_2-$MgCl_2$-Na_2SO_4-K_2SO_4-$MgSO_4$-H_2O

bi: bischofite ($MgCl_2·6H_2O$), ca: carnallite ($KMgCl_3·6H_2O$)
gs: glaserite ($K_3Na(SO_4)_2$), ha: halite (NaCl)
ka: kainite ($KMgClSO_4·11/4\ H_2O$), ks: kieserite ($MgSO_4·H_2O$)
le: leonite ($K_2Mg(SO_4)_2·4H_2O$), pc: picromerite ($K_2Mg(SO_4)_2·6H_2O$)
sy: sylvite (KCl)

25°C
Saturation with respect to sy + ha and ca + ha
Water values as a function of the total concentration of Mg^{2+}

halite and	$\{Mg^{2+}\}$	W_T
F: gs + sy	0	1457.7
M: gs + pc + sy	35.2	1351.4
N: le + pc + sy	39.7	1331.6
P: ka + le + sy	57.2	1273.9
Q: ca + ka + sy	73.1	1200.5
R: ca + ka + ks	86.8	1103.8
Z: bi + ca + ks	105.3	941.6

$$\{Mg^{2+}\} = \{MgCl_2\} + \{MgSO_4\} \qquad W_T = \frac{10^5}{C_T}$$

$$C_T = \{Na_2^{2+}\} + \{K_2^{2+}\} + \{Mg^{2+}\} = \{Cl_2^{2-}\} + \{SO_4^{2-}\}$$

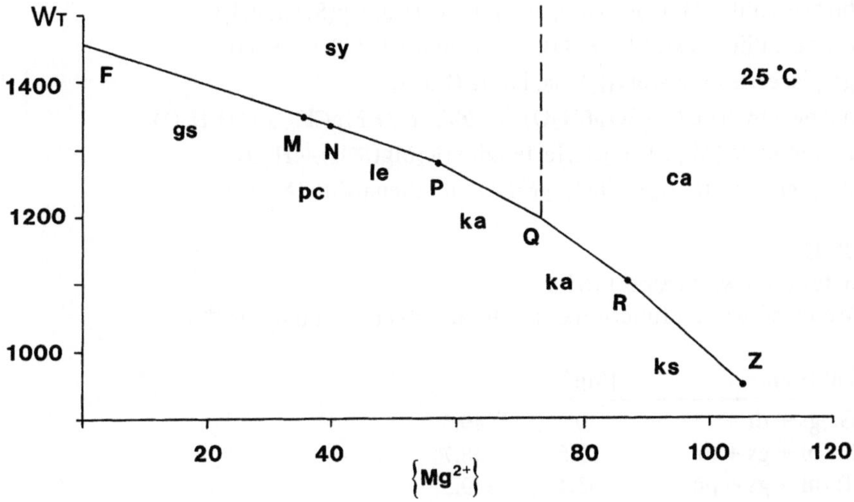
WT
1400
1200
1000
20
40
80
100
120
{Mg2+}
25 °C
sy
F
gs
M
N
le
P
pc
ka
Q
ca
ka
R
ks
Z

III.5
Na_2Cl_2-K_2Cl_2-$MgCl_2$-Na_2SO_4-K_2SO_4-$MgSO_4$-H_2O

bi: bischofite ($MgCl_2 \cdot 6H_2O$), bl: bloedite ($Na_2Mg(SO_4)_2 \cdot 4H_2O$)
ca: carnallite ($KMgCl_3 \cdot 6H_2O$), ep: epsomite ($MgSO_4 \cdot 7H_2O$)
gs: glaserite ($K_3Na(SO_4)_2$), ha: halite (NaCl)
hx: hexahydrite ($MgSO_4 \cdot 6H_2O$), ka: kainite ($KMgClSO_4 \cdot 11/4\ H_2O$)
ks: kieserite ($MgSO_4 \cdot H_2O$), le: leonite ($K_2Mg(SO_4)_2 \cdot 4H_2O$)
pc: picromerite ($K_2Mg(SO_4)_2 \cdot 6H_2O$), th: thenardite ($Na_2SO_4$)

25°C
Saturation with respect to ha
Water values as a function of the total concentration of Mg^{2+}

halite and	$\{Mg^{2+}\}$	W_T
G: gs + th	0	1449.3
S: bl + gs + th	19.5	1338.7
T: bl + gs + pc	33.4	1349.5
U: bl + le + pc	36.3	1340.5
V: bl + ep + le	62.3	1253.1
W: ep + ka + le	63.6	1234.6
X: ep + hx + ka	73.8	1193.3
Y: hx + ka + ks	83.1	1137.7
R: ca + ka + ks	86.8	1103.8
Z: bi + ca + ks	105.3	941.6

$$\{Mg^{2+}\} = \{MgCl_2\} + \{MgSO_4\} \qquad W_T = \frac{10^5}{C_T}$$

$$C_T = \{Na_2^{2+}\} + \{K_2^{2+}\} + \{Mg^{2+}\} = \{Cl_2^{2-}\} + \{SO_4^{2-}\}$$

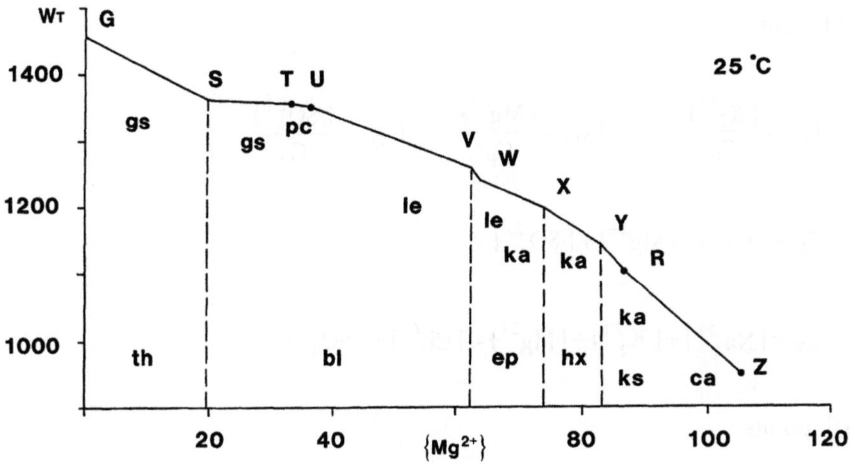
WT
G
1400
1200
1000
S
T
U
25 °C
gs
gs
pc
V
W
X
le
le
ka
ka
Y
R
ka
th
bl
ep
hx
ks
ca
Z
20
40
{Mg2+}
80
100
120

III.5
Na_2Cl_2-K_2Cl_2-$MgCl_2$-Na_2SO_4-K_2SO_4-$MgSO_4$-H_2O

Solutions

$$F_{K_2} = \frac{\{K_2^{2+}\}}{C_P} \qquad F_{Mg} = \frac{\{Mg^{2+}\}}{C_P} \qquad F_{SO_4} = \frac{\{SO_4^{2-}\}}{C_P}$$

$$C_P = \{K_2^{2+}\} + \{Mg^{2+}\} + \{SO_4^{2-}\}$$

$$C_T = \{Na_2^{2+}\} + \{K_2^{2+}\} + \{Mg^{2+}\} = \{Cl_2^{2-}\} + \{SO_4^{2-}\}$$

Salt points

		F_{K_2}	F_{Mg}	F_{SO_4}
bi: bischofite	$MgCl_2 \cdot 6H_2O$	0	1	0
bl: bloedite	$Na_2Mg(SO_4)_2 \cdot 4H_2O$	0	1/3	2/3
ca: carnallite	$KMgCl_3 \cdot 6H_2O$	1/3	2/3	0
da: d'ansite	$Na_{21}MgCl_3(SO_4)_{10}$	0	1/11	10/11
ep: epsomite	$MgSO_4 \cdot 7H_2O$	0	1/2	1/2
gs: glaserite	$K_3Na(SO_4)_2$	3/7	0	4/7
hx: hexahydrite	$MgSO_4 \cdot 6H_2O$	0	1/2	1/2
ka: kainite	$KMgClSO_4 \cdot 11/4\ H_2O$	1/5	2/5	2/5
ks: kieserite	$MgSO_4 \cdot H_2O$	0	1/2	1/2
lg: langbeinite	$K_2Mg_2(SO_4)_3$	1/6	2/6	3/6
le: leonite	$K_2Mg(SO_4)_2 \cdot 4H_2O$	1/4	1/4	2/4
lw: loeweite	$Na_{12}Mg_7(SO_4)_{13} \cdot 15H_2O$	0	7/20	13/20
mi: mirabilite	$Na_2SO_4 \cdot 10H_2O$	0	0	1
pc: picromerite	$K_2Mg(SO_4)_2 \cdot 6H_2O$	1/4	1/4	2/4
sy: sylvite	KCl	1	0	0
th: thenardite	Na_2SO_4	0	0	1
vh: vanthoffite	$Na_6Mg(SO_4)_4$	0	1/5	4/5
sw: seawater		0.057	0.616	0.327

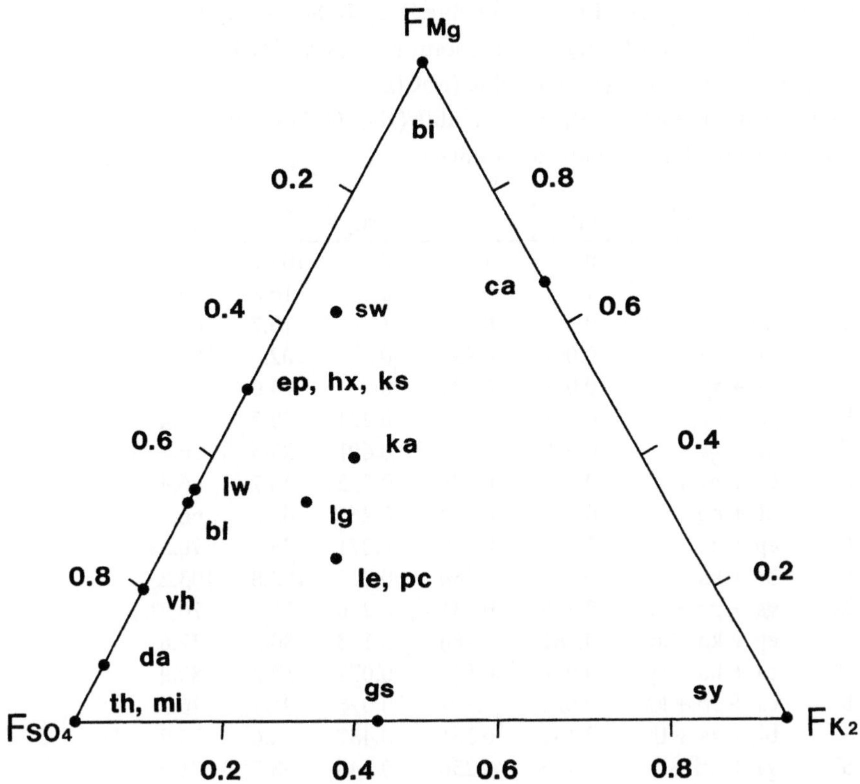

FMg
bi
0.2
0.8
ca
0.4
sw
0.6
ep, hx, ks
0.6
ka
0.4
lw
lg
bl
0.8
le, pc
0.2
vh
da
gs
sy
th, mi
FSO4
FK2
0.2
0.4
0.6
0.8

III.5
Na_2Cl_2-K_2Cl_2-$MgCl_2$-Na_2SO_4-K_2SO_4-$MgSO_4$-H_2O

bi: bischofite ($MgCl_2 \cdot 6H_2O$), bl: bloedite ($Na_2Mg(SO_4)_2 \cdot 4H_2O$)
ca: carnallite ($KMgCl_3 \cdot 6H_2O$), ep: epsomite ($MgSO_4 \cdot 7H_2O$)
gs: glaserite ($K_3Na(SO_4)_2$), ha: halite (NaCl)
ks: kieserite ($MgSO_4 \cdot H_2O$), mi: mirabilite ($Na_2SO_4 \cdot 10H_2O$)
sy: sylvite (KCl), th: thenardite (Na_2SO_4)

15°C	halite and	F_{K_2}	F_{Mg}	F_{SO_4}	C_P	C_T
A	bi	0	1	0	101.7	102.3
B	sy	1	0	0	16.8	63.9
C	mi	0	0	1	10.7	61.1
D	bi + ca	0.002	0.998	0	102.3	103.1
E	ca + sy	0.069	0.931	0	74.0	78.1
F	gs + sy	0.779	0	0.221	22.2	67.4
G	mi + gs	0.399	0	0.601	23.3	66.7
H	bl + mi + th	0	0.428	0.572	36.7	66.4
I	bl + ep	0	0.734	0.266	62.0	66.7
J	ep + ks	0	0.929	0.071	95.4	90.5
L	bi + ks	0	0.988	0.012	103.8	103.2
M	gs + pc + sy	0.249	0.531	0.220	55.9	70.9
P	ep + ka + sy	0.081	0.786	0.133	80.5	77.6
Q	ca + ka + sy	0.057	0.872	0.071	82.2	80.9
R	ca + ep + ka	0.021	0.901	0.078	91.1	86.7
S	bl + gs + th	0.132	0.381	0.487	54.6	75.9
S'	gs + mi + th	0.196	0.256	0.548	38.7	71.9
T	bl + gs + pc	0.200	0.516	0.284	56.0	71.4
V	bl + ep + pc	0.118	0.679	0.203	72.3	74.0
W	ep + pc + sy	0.103	0.729	0.168	77.9	75.7
Y	ca + ep + ks	0.009	0.921	0.070	6.2	91.3
Z	bi + ca + ks	0.002	0.986	0.012	104.6	103.9

H: 16.2°C

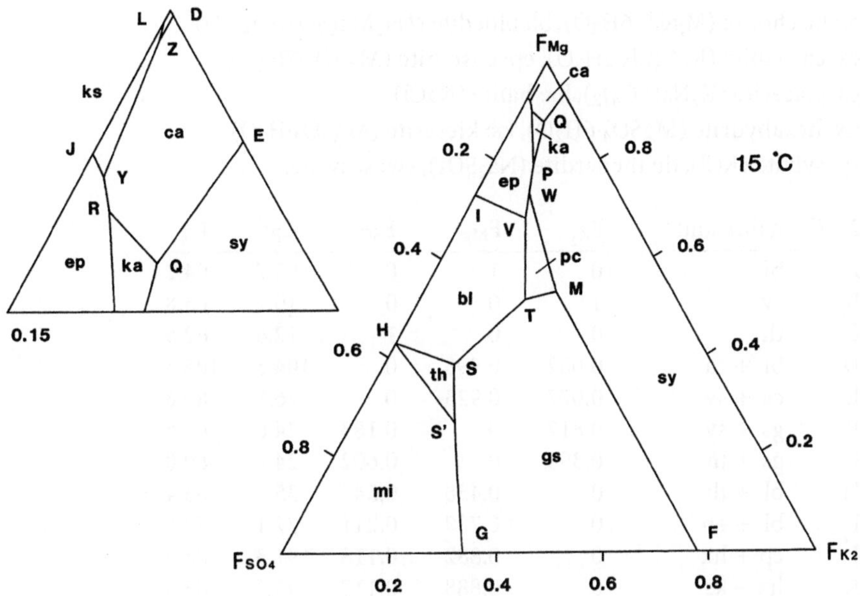
L
D
Z
ks
ca
J
E
Y
R
sy
ep
ka
Q
0.15
FMg
ca
Q
ka
0.2
0.8
15 °C
P
ep
W
I
V
0.4
pc
0.6
M
bl
T
H
0.6
0.4
th
S
sy
S'
0.8
0.2
gs
mi
G
F
FSO4
FK2
0.2
0.4
0.6
0.8

III.5
Na_2Cl_2-K_2Cl_2-$MgCl_2$-Na_2SO_4-K_2SO_4-$MgSO_4$-H_2O

bi: bischofite ($MgCl_2{\cdot}6H_2O$), bl: bloedite ($Na_2Mg(SO_4)_2{\cdot}4H_2O$)
ca: carnallite ($KMgCl_3{\cdot}6H_2O$), ep: epsomite ($MgSO_4{\cdot}7H_2O$)
gs: glaserite ($K_3Na(SO_4)_2$), ha: halite (NaCl)
hx: hexahydrite ($MgSO_4{\cdot}6H_2O$), ks: kieserite ($MgSO_4{\cdot}H_2O$)
sy: sylvite (KCl), th: thenardite (Na_2SO_4), sw: seawater

25°C	halite and	F_{K_2}	F_{Mg}	F_{SO_4}	C_P	C_T
A	bi	0	1	0	103.7	104.2
B	sy	1	0	0	19.7	65.8
C	th	0	0	1	12.6	62.5
D	bi + ca	0.002	0.998	0	104.5	105.3
E	ca + sy	0.077	0.923	0	76.5	80.6
F	gs + sy	0.817	0	0.183	24.6	68.6
G	gs + th	0.398	0	0.602	24.9	69.0
H	bl + th	0	0.458	0.542	35.4	66.4
I	bl + ep	0	0.789	0.211	77.1	72.6
J	ep + hx	0	0.882	0.118	91.3	84.5
K	hx + ks	0	0.888	0.112	93.6	85.8
L	bi + ks	0	0.990	0.010	105.4	105.1
M	gs + pc + sy	0.228	0.553	0.219	63.6	74.4
N	le + pc + sy	0.202	0.589	0.209	67.4	75.1
P	ka + le + sy	0.120	0.707	0.173	80.9	78.5
Q	ca + ka + sy	0.068	0.872	0.060	83.8	83.3
R	ca + ka + ks	0.015	0.919	0.066	94.4	90.6
S	bl + gs + th	0.179	0.371	0.450	52.5	74.7
T	bl + gs + pc	0.181	0.518	0.301	64.5	74.1
U	bl + le + pc	0.171	0.543	0.286	66.8	74.6
V	bl + ep + le	0.072	0.705	0.223	88.4	79.8
W	ep + ka + le	0.077	0.715	0.208	88.9	81.0
X	ep + hx + ka	0.034	0.838	0.128	88.1	83.8
Y	hx + ka + ks	0.017	0.868	0.115	95.7	87.9
Z	bi + ca + ks	0.002	0.988	0.010	106.6	106.2

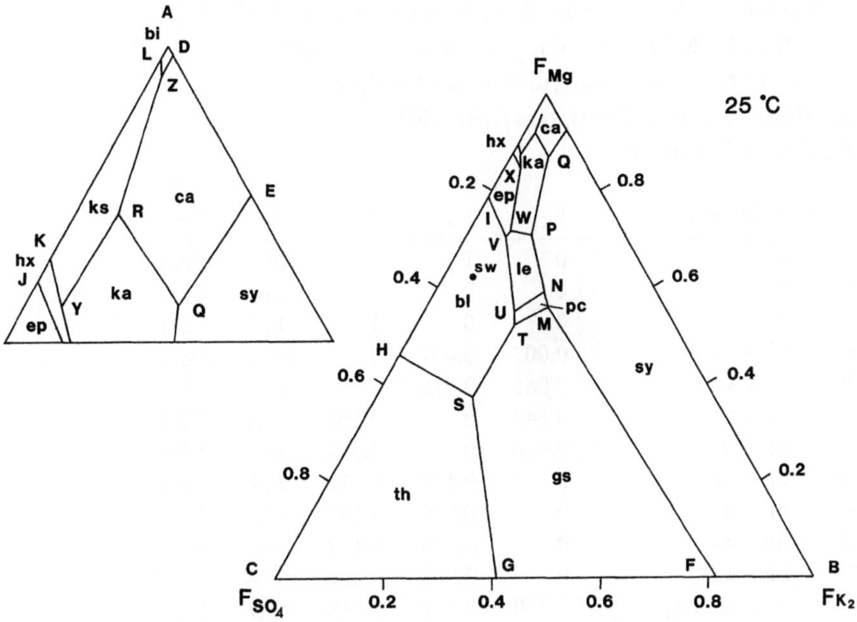
A
bi
L
D
Z
ca
E
ks
R
K
hx
J
ka
sy
Y
Q
ep
F_Mg
25 °C
hx
ca
ka
Q
X
0.2
ep
0.8
W
I
P
V
0.4
sw
le
0.6
N
bl
U
pc
M
T
H
sy
0.6
0.4
S
0.8
gs
0.2
th
C
G
F
B
F_SO4
0.2
0.4
0.6
0.8
F_K2

III.5
Na_2Cl_2-K_2Cl_2-$MgCl_2$-Na_2SO_4-K_2SO_4-$MgSO_4$-H_2O

bi: bischofite ($MgCl_2 \cdot 6H_2O$), bl: bloedite ($Na_2Mg(SO_4)_2 \cdot 4H_2O$)
ca: carnallite ($KMgCl_3 \cdot 6H_2O$), gs: glaserite ($K_3Na(SO_4)_2$)
ha: halite (NaCl), hx: hexahydrite ($MgSO_4 \cdot 6H_2O$)
ks: kieserite ($MgSO_4 \cdot H_2O$), sy: sylvite (KCl)
th: thenardite (Na_2SO_4)

35°C	halite and	F_{K_2}	F_{Mg}	F_{SO_4}	C_P	C_T
A	bi	0	1	0	106.3	106.8
B	sy	1	0	0	22.6	67.8
C	th	0	0	1	11.2	62.2
D	bi + ca	0.003	0.997	0	107.3	108.1
E	ca + sy	0.085	0.915	0	79.2	83.3
F	gs + sy	0.843	0	0.157	27.4	70.4
G	gs + th	0.456	0	0.544	25.2	69.6
H	bl + th	0	0.488	0.512	34.8	66.4
I	bl + hx	0	0.800	0.200	92.0	81.2
K	hx+ ks	0	0.838	0.162	94.3	84.3
L	bi + ks	0	0.993	0.007	107.4	107.3
M	gs + le + sy	0.250	0.537	0.213	65.7	75.9
P	ka + le + sy	0.156	0.663	0.181	80.6	80.0
Q	ca + ka + sy	0.077	0.872	0.051	86.0	86.1
R	ca + ka + ks	0.025	0.921	0.054	93.5	91.1
S	bl + gs + th	0.225	0.358	0.417	50.6	74.1
T	bl + gs + le	0.203	0.504	0.293	64.9	75.8
V	bl + ka + le	0.112	0.663	0.225	85.4	80.6
Y	hx+ ka + ks + lw	0.020	0.797	0.183	98.0	85.5
Y'	bl + hx + ka	0.030	0.764	0.206	96.7	83.9
Z	bi + ca + ks	0.003	0.989	0.008	109.0	108.9

Y: 34.5°C

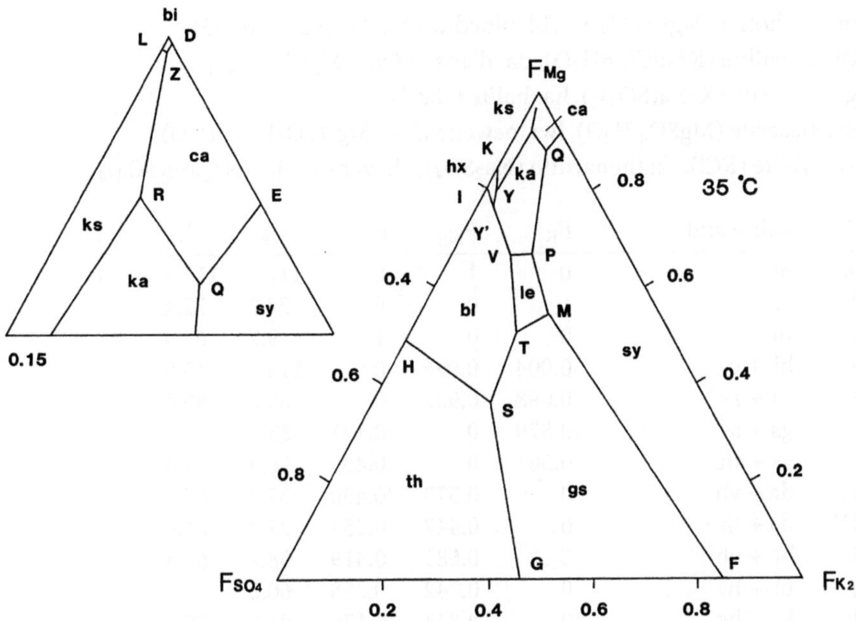
bi
L
D
Z
ca
ks
R
E
ka
Q
sy
0.15
F$_{Mg}$
ks
ca
K
Q
hx
ka
0.8
35 °C
I
Y
Y'
V
P
0.4
0.6
le
bl
M
T
sy
0.6
H
0.4
S
0.8
th
gs
0.2
G
F
F$_{SO_4}$
F$_{K_2}$
0.2
0.4
0.6
0.8

III.5
Na_2Cl_2-K_2Cl_2-$MgCl_2$-Na_2SO_4-K_2SO_4-$MgSO_4$-H_2O

bi: bischofite ($MgCl_2{\cdot}6H_2O$), bl: bloedite ($Na_2Mg(SO_4)_2{\cdot}4H_2O$)
ca: carnallite ($KMgCl_3{\cdot}6H_2O$), da: d'ansite ($Na_{21}MgCl_3(SO_4)_{10}$)
gs: glaserite ($K_3Na(SO_4)_2$), ha: halite (NaCl)
ks: kieserite ($MgSO_4{\cdot}H_2O$), lw: loeweite ($Na_{12}Mg_7(SO_4)_{13}{\cdot}15H_2O$)
sy: sylvite (KCl), th: thenardite (Na_2SO_4), vh: vanthoffite ($Na_6Mg(SO_4)_4$)

55°C	halite and	F_{K_2}	F_{Mg}	F_{SO_4}	C_P	C_T
A	bi	0	1	0	112.9	113.4
B	sy	1	0	0	28.7	72.4
C	th	0	0	1	9.2	62.3
D	bi + ca	0.004	0.996	0	114.1	115.0
E	ca + sy	0.098	0.902	0	85.6	89.8
F	gs + sy	0.879	0	0.121	33.1	74.3
G	gs + th	0.561	0	0.439	26.9	71.3
H	da + vh	0	0.570	0.430	37.2	67.2
H'	da + th	0	0.447	0.553	27.3	66.6
I	bl + vh	0	0.581	0.419	38.7	67.4
J	bl + lw	0	0.742	0.258	60.0	70.1
K	ks + lw	0	0.824	0.176	83.0	79.4
L	bi + ks	0	0.996	0.004	113.2	113.5
M	gs + le + sy	0.294	0.495	0.211	74.1	83.1
N	ka + le + lg + sy	0.235	0.569	0.196	81.7	84.9
Q	ca + ka + sy	0.095	0.868	0.037	91.5	92.5
R	ca + ka + ks	0.064	0.899	0.037	94.0	94.0
S	da + gs + vh	0.275	0.370	0.355	53.5	76.3
S'	da + gs + th	0.347	0.249	0.404	40.6	73.5
T	bl + gs + vh	0.247	0.410	0.343	59.5	77.6
U	bl + gs + le	0.231	0.492	0.277	75.4	81.9
V	bl + le + lw	0.218	0.507	0.275	76.1	82.2
W	le + lg + lw	0.187	0.570	0.243	82.6	83.7
X	ka + lg + lw	0.115	0.672	0.213	86.5	83.1
Y	ka + ks + lw	0.045	0.777	0.178	92.6	85.5
Z	bi + ca + ks	0.004	0.991	0.005	115.2	115.5

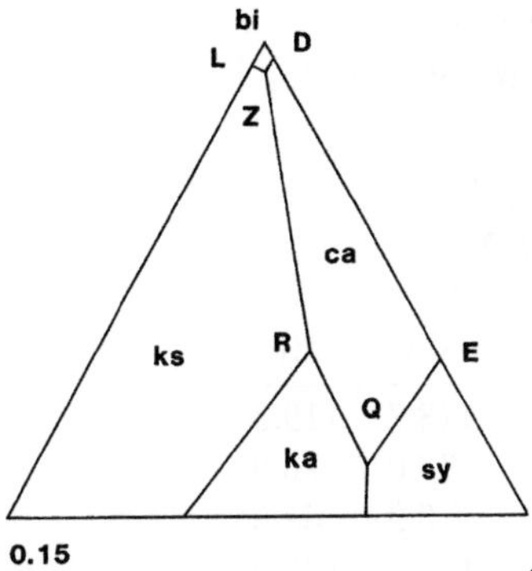
bi
L
D
Z
ca
ks
R
E
Q
ka
sy
0.15

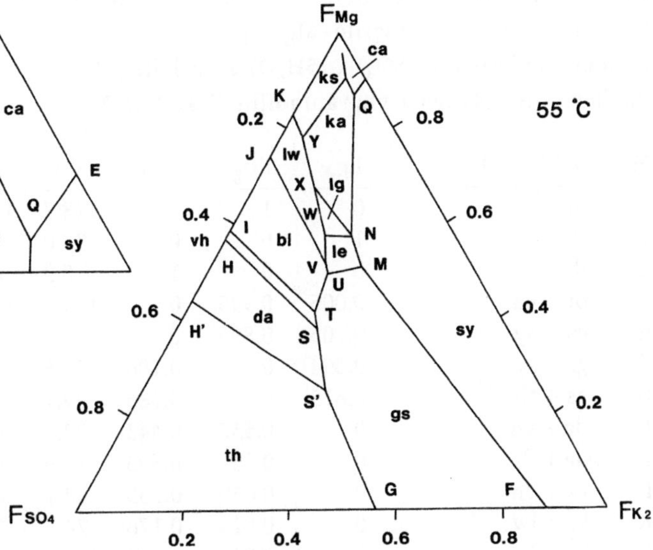
FMg
ca
ks
K
Q
0.2
ka
0.8
55 °C
Y
J
lw
X
lg
0.4
W
0.6
I
N
vh
bl
le
H
V
M
U
0.6
T
0.4
da
H'
S
sy
0.8
S'
0.2
gs
th
G
F
FSO4
FK2
0.2
0.4
0.6
0.8

III.5
Na_2Cl_2-K_2Cl_2-$MgCl_2$-Na_2SO_4-K_2SO_4-$MgSO_4$-H_2O

bi: bischofite ($MgCl_2{\cdot}6H_2O$), ca: carnallite ($KMgCl_3{\cdot}6H_2O$)
da: d'ansite ($Na_{21}MgCl_3(SO_4)_{10}$), gs: glaserite ($K_3Na(SO_4)_2$)
ha: halite (NaCl), ks: kieserite ($MgSO_4{\cdot}H_2O$)
lw: loeweite ($Na_{12}Mg_7(SO_4)_{13}{\cdot}15H_2O$), sy: sylvite (KCl)
th: thenardite (Na_2SO_4), vh: vanthoffite ($Na_6Mg(SO_4)_4$)

69°C	halite and	F_{K_2}	F_{Mg}	F_{SO_4}	C_P	C_T
A	bi	0	1	0	118.7	119.3
B	sy	1	0	0	33.0	76.0
C	th	0	0	1	8.3	62.8
D	bi + ca	0.005	0.995	0	119.9	120.9
E	ca + sy	0.105	0.895	0	90.7	95.1
F	gs + sy	0.894	0	0.106	37.6	77.8
G	gs + th	0.615	0	0.385	29.1	73.3
H	da + vh	0	0.557	0.443	27.3	66.1
H'	da + th	0	0.307	0.693	17.9	66.7
I	lw + vh	0	0.665	0.335	42.1	68.3
K	ks + lw	0	0.824	0.176	74.3	76.5
L	bi + ks	0	0.996	0.004	118.9	119.2
M	gs + lg + sy	0.394	0.419	0.187	68.9	84.7
Q	ca + ka + ks + sy	0.105	0.864	0.031	96.4	97.7
R	ka + lg + sy	0.196	0.711	0.093	83.7	87.7
S	da + gs + vh	0.379	0.289	0.332	46.7	76.9
S'	da + gs + th	0.484	0.120	0.396	34.3	74.3
V	gs + lw + vh	0.329	0.409	0.262	63.6	81.6
W	gs + lg + lw	0.324	0.415	0.261	64.4	82.0
X	ks + lg + lw	0.131	0.718	0.151	88.8	87.8
Y	ka + ks + lg	0.133	0.722	0.145	90.4	88.8
Z	bi + ca + ks	0.005	0.991	0.004	120.6	121.1

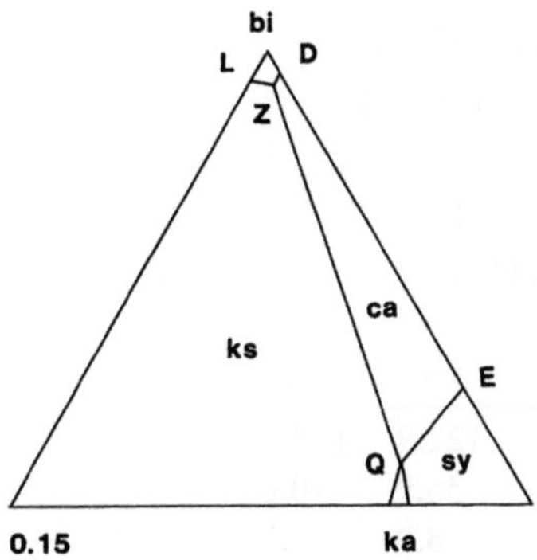

bi
L
D
Z
ca
ks
E
Q
sy
0.15
ka

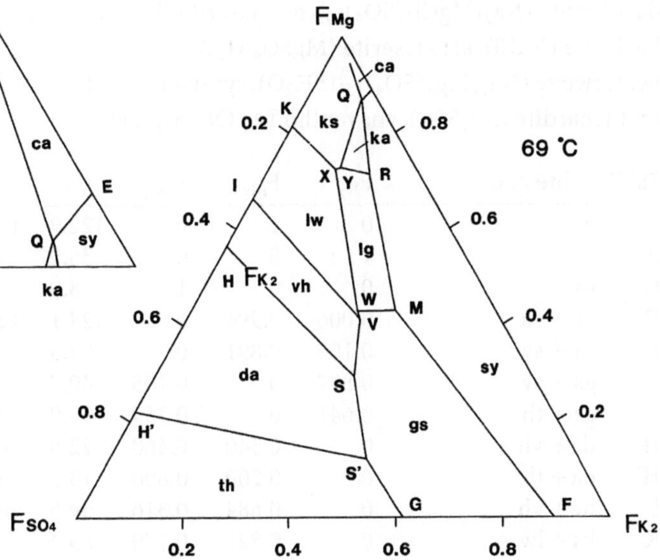

FMg
ca
Q
K
0.2
ks
0.8
ka
69 °C
X
Y
R
I
0.4
lw
0.6
lg
H
FK2
vh
W
M
V
0.6
0.4
sy
da
S
0.8
0.2
gs
H'
S'
th
G
F
FSO4
FK2
0.2
0.4
0.6
0.8

III.5
Na_2Cl_2-K_2Cl_2-$MgCl_2$-Na_2SO_4-K_2SO_4-$MgSO_4$-H_2O

bi: bischofite ($MgCl_2 \cdot 6H_2O$), ca: carnallite ($KMgCl_3 \cdot 6H_2O$)
da: d'ansite ($Na_{21}MgCl_3(SO_4)_{10}$), gs: glaserite ($K_3Na(SO_4)_2$)
ha: halite (NaCl), ks: kieserite ($MgSO_4 \cdot H_2O$)
lw: loeweite ($Na_{12}Mg_7(SO_4)_{13} \cdot 15H_2O$), sy: sylvite (KCl)
th: thenardite (Na_2SO_4), vh: vanthoffite ($Na_6Mg(SO_4)_4$)

78°C	halite and	F_{K_2}	F_{Mg}	F_{SO_4}	C_P	C_T
A	bi	0	1	0	122.9	123.6
B	sy	1	0	0	35.9	78.6
C	th	0	0	1	8.0	63.3
D	bi + ca	0.006	0.994	0	124.1	125.1
E	ca + sy	0.109	0.891	0	94.3	98.8
F	gs + sy	0.897	0	0.103	40.7	80.3
G	gs + th	0.641	0	0.359	30.9	74.8
H	da + vh	0	0.540	0.460	22.6	65.9
H'	da + th	0	0.200	0.800	13.5	66.4
I	lw + vh	0	0.684	0.316	39.5	68.3
K	ks + lw	0	0.821	0.179	69.4	75.5
L	bi + ks	0	0.996	0.004	123.1	123.5
M	gs + lg + sy	0.476	0.354	0.170	65.3	86.1
Q	ca + ks + sy	0.110	0.865	0.025	98.9	100.6
R	ka + ks + lg + sy	0.161	0.785	0.054	90.7	93.3
S	da + gs + vh	0.458	0.230	0.312	44.3	78.2
S'	da + gs + th	0.564	0.054	0.382	33.0	75.5
V	gs + lg + vh	0.392	0.351	0.257	57.6	81.2
W	lg + lw + vh	0.328	0.416	0.256	59.1	80.8
Y	ks + lg + lw	0.132	0.703	0.165	79.3	83.8
Z	bi + ca + ks	0.006	0.990	0.004	124.8	125.4

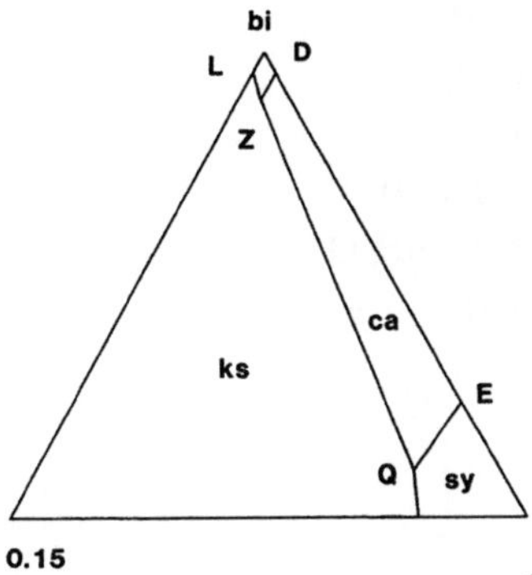
bi
L
D
Z
ca
ks
E
Q
sy
0.15

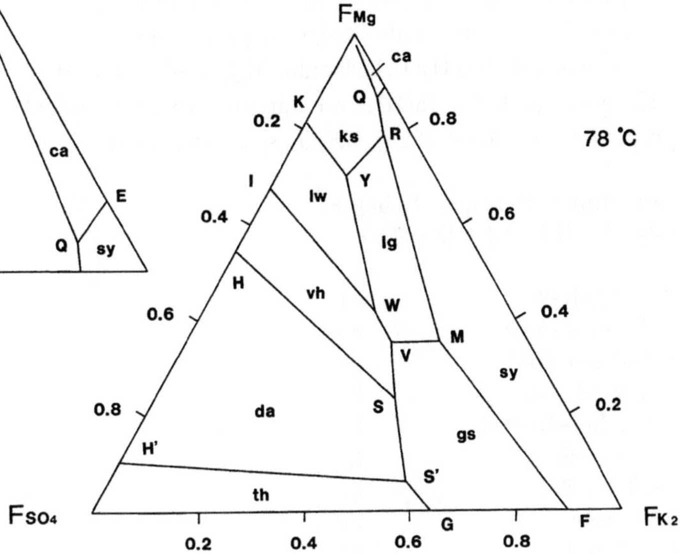
FMg
ca
Q
K
0.2
0.8
ks
R
78 °C
I
Y
lw
0.4
0.6
lg
H
vh
W
0.6
0.4
M
V
sy
0.8
da
S
0.2
gs
H'
S'
th
FSO4
G
F
FK2
0.2
0.4
0.6
0.8

III.5
Na_2Cl_2-K_2Cl_2-$MgCl_2$-Na_2SO_4-K_2SO_4-$MgSO_4$-H_2O

bi: bischofite ($MgCl_2 \cdot 6H_2O$), ca: carnallite ($KMgCl_3 \cdot 6H_2O$)
ep: epsomite ($MgSO_4 \cdot 7H_2O$), gs: glaserite ($K_3Na(SO_4)_2$)
ha: halite (NaCl), ka: kainite ($KMgClSO_4 \cdot 11/4\ H_2O$)
ks: kieserite ($MgSO_4 \cdot H_2O$), le: leonite ($K_2Mg(SO_4)_2 \cdot 4H_2O$)
lg: langbeinite ($K_2Mg_2(SO_4)_3$), mi: mirabilite ($Na_2SO_4 \cdot 10H_2O$)
pc: picromerite ($K_2Mg(SO_4)_2 \cdot 6H_2O$), sy: sylvite (KCl)

Saturation with respect to halite
$\{Mg^{2+}\} = \{MgCl_2\} + \{MgSO_4\}$

d–e–f–g–h–9–2:	ca
c–d–e–4–2–5–3:	ep
a–1–6–8–j–100°:	gs
2–4–9–10–7–5:	ka
e–f–g–h–i–10–9–4:	ks
3–5–7–8–6:	le
7–10–i–j–8:	lg
0°–a–1–b:	mi
b–c–3–6–1:	pc
0°–b–c–d–2–9–h–i–j–100°:	sy

	halite and		
1–a:	mi + gs + sy	4–9:	ca + ka + ks
1–b:	mi + pc + sy	5–7:	ka + le + sy
1–6:	gs + pc + sy	6–8:	gs + le + sy
2–d:	ca + ep + sy	7–8:	le + lg + sy
2–4:	ca + ep + ka	7–10:	ka + lg + sy
2–5:	ep + ka + sy	8–j:	gs + lg + sy
2–9:	ca + ka + sy	9–h:	ca + ks + sy
3–c:	ep + pc + sy	9–10:	ka + ks + sy
3–5:	ep + le + sy	10–i:	ks + lg + sy
3–6:	le + pc + sy	f–g:	bi + ca + ks
4–e:	ca + ep + ks		

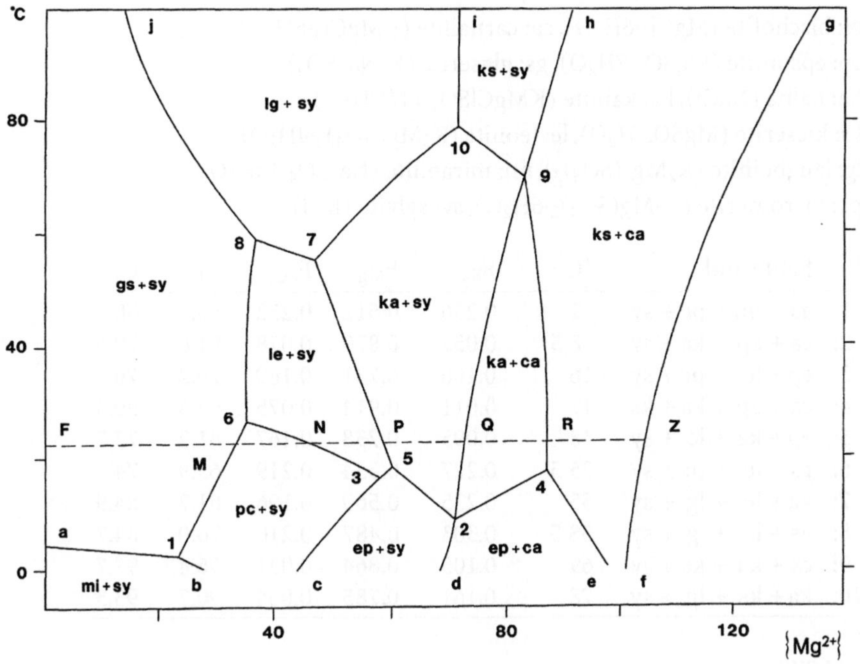
°C
80
40
0
40
80
120
{Mg2+}
j
i
h
g
ks+sy
lg + sy
10
9
8
7
ks+ca
gs+sy
ka+sy
le+sy
ka+ca
F
6
N
P
Q
R
Z
M
5
3
4
pc+sy
2
a
1
ep+sy
ep+ca
mi+sy
b
c
d
e
f

III.5
Na_2Cl_2-K_2Cl_2-$MgCl_2$-Na_2SO_4-K_2SO_4-$MgSO_4$-H_2O

bi: bischofite ($MgCl_2 \cdot 6H_2O$), ca: carnallite ($KMgCl_3 \cdot 6H_2O$)
ep: epsomite ($MgSO_4 \cdot 7H_2O$), gs: glaserite ($K_3Na(SO_4)_2$)
ha: halite (NaCl), ka: kainite ($KMgClSO_4 \cdot 11/4\ H_2O$)
ks: kieserite ($MgSO_4 \cdot H_2O$), le: leonite ($K_2Mg(SO_4)_2 \cdot 4H_2O$)
lg: langbeinite ($K_2Mg_2(SO_4)_3$), mi: mirabilite ($Na_2SO_4 \cdot 10H_2O$)
pc: picromerite ($K_2Mg(SO_4)_2 \cdot 6H_2O$), sy: sylvite (KCl)

	halite and	°C	F_{K_2}	F_{Mg}	F_{SO_4}	C_P	C_T
1:	gs + mi + pc + sy	2	0.236	0.512	0.252	46.1	68.6
2:	ca + ep + ka + sy	8.5	0.052	0.870	0.078	81.6	79.6
3:	ep + le + pc + sy	16	0.100	0.731	0.169	79.2	76.3
4:	ca + ep + ka + ks	17	0.011	0.914	0.075	95.5	90.4
5:	ep + ka + le + sy	18	0.095	0.738	0.167	81.2	77.5
6:	gs + le + pc + sy	25.5	0.227	0.554	0.219	63.9	74.1
7:	ka + le + lg + sy	55	0.235	0.569	0.196	81.7	84.9
8:	gs + le + lg + sy	58.5	0.303	0.487	0.210	76.0	84.7
9:	ca + ka + ks + sy	69	0.105	0.864	0.031	96.4	97.7
10:	ka + ks + lg + sy	78	0.161	0.785	0.054	90.7	93.3

SALSYS
Input: temperature. Output: {Na_2Cl_2}, {K_2Cl_2}, {$MgCl_2$}, {$MgSO_4$}

	halite and	°C		
1–6 (M):	gs + pc + sy	2.0	-	25.5
2–d (Q):	ca + ep + sy	8.5	-	0
2–9 (Q):	ca + ka + sy	8.5	-	69.0
4–9 (R):	ca + ka + ks	17.0	-	69.0
5–7 (P):	ka + le + sy	18.0	-	55.0
6–8 (M):	gs + le + sy	25.5	-	58.5
8–j (M):	gs + lg + sy	58.5	-	90.0
9–h (Q):	ca + ks + sy	69.0	-	90.0
f–g (Z):	bi + ca + ks	15.0	-	78.0

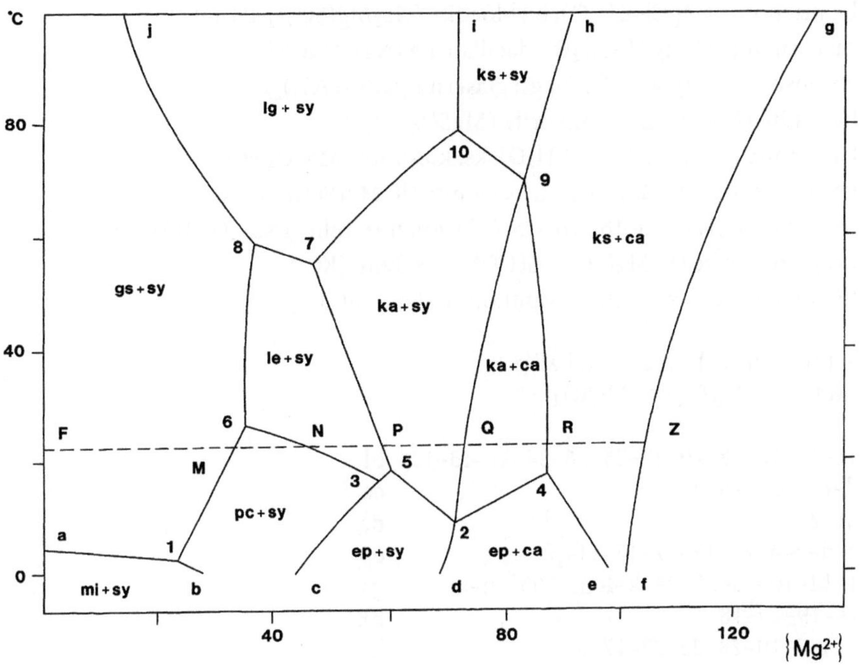
°C
80
40
0
40
80
120
{Mg2+}
j
i
h
g
ks+sy
lg + sy
10
9
8
7
ks+ca
gs+sy
ka+sy
le+sy
ka+ca
6
F
N
P
Q
R
Z
M
5
3
4
pc+sy
a
1
2
ep+sy
ep+ca
mi+sy
b
c
d
e
f

III.5 Na_2Cl_2-K_2Cl_2-$MgCl_2$-Na_2SO_4-K_2SO_4-$MgSO_4$-H_2O

bi: bischofite ($MgCl_2 \cdot 6H_2O$), bl: bloedite ($Na_2Mg(SO_4)_2 \cdot 4H_2O$)
ca: carnallite ($KMgCl_3 \cdot 6H_2O$), da: d'ansite ($Na_{21}MgCl_3(SO_4)_{10}$)
ep: epsomite ($MgSO_4 \cdot 7H_2O$), gs: glaserite ($K_3Na(SO_4)_2$)
ha: halite (NaCl), hx: hexahydrite ($MgSO_4 \cdot 6H_2O$)
ka: kainite ($KMgClSO_4 \cdot 11/4\ H_2O$), ks: kieserite ($MgSO_4 \cdot H_2O$)
lg: langbeinite ($K_2Mg_2(SO_4)_3$), le: leonite ($K_2Mg(SO_4)_2 \cdot 4H_2O$)
lw: loeweite ($Na_{12}Mg_7(SO_4)_{13} \cdot 15H_2O$), mi: mirabilite ($Na_2SO_4 \cdot 10H_2O$)
pc: picromerite ($K_2Mg(SO_4)_2 \cdot 6H_2O$), sy: sylvite (KCl)
th: thenardite (Na_2SO_4), vh: vanthoffite ($Na_6Mg(SO_4)_4$)

Saturation with respect to halite
$\{Mg^{2+}\} = \{MgCl_2\} + \{MgSO_4\}$

11–15–17–18–20–22–25–26–24–21–13–12:	bl
d–e–f–g–h–9–4–2:	ca
21–24–l–m:	da
c–d–e–4–14–18–17–15–11–p:	ep
1–12–16–25–27–29–k–l–m–100°–n–a:	gs
14–19–20–18:	hx
2–4–9–10–28–23–22–17–5:	ka
e–f–g–h–i–30–28–19–14–4:	ks
3–5–17–22–23–27–25–16–15:	le
23–28–10–i–30–k–29–27:	lg
19–28–30–29–26–25–22–20:	lw
0°–b–p–11–12–13–n–a:	mi
c–3–15–16–12–1–b–p:	pc
0°–b–1–a and c–d–2–5–3 and 9–h–i–10:	sy
13–21–m–100°–n:	th
24–26–29–30–k–l:	vh

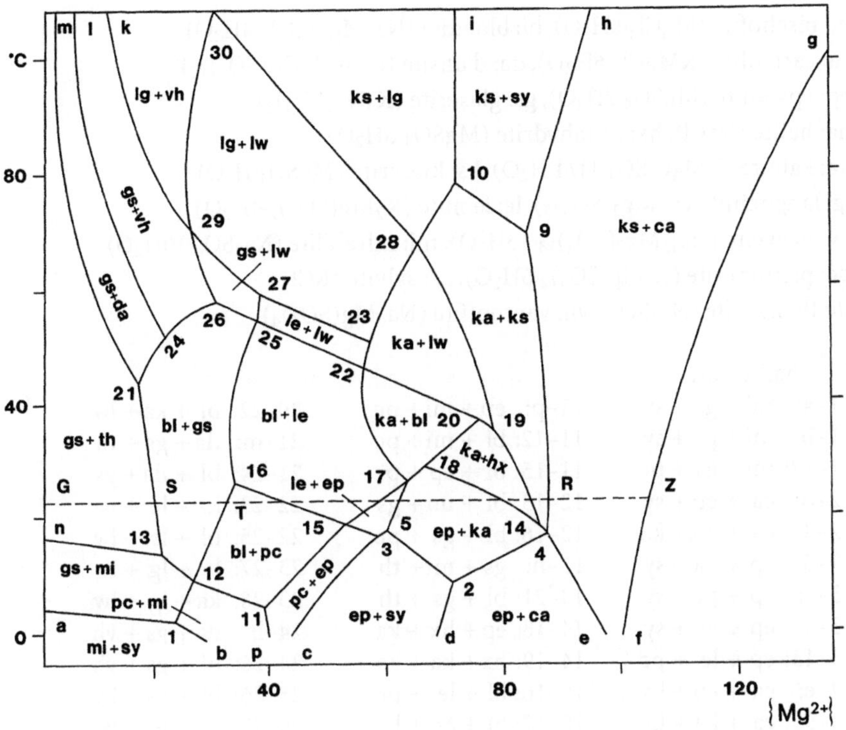
°C
80
40
0
40
80
120
{Mg2+{
m
l
k
i
h
g
30
ks + lg
ks + sy
lg + vh
lg + lw
10
gs+vh
29
9
ks + ca
28
gs + lw
27
gs+ap
26
23
le + lw
ka + ks
24
25
ka + lw
22
21
bl + le
ka + bl
20
19
bl + gs
gs + th
ka+hx
18
16
le + ep
17
G
S
R
Z
T
n
13
15
5
ep + ka
14
3
4
bl + pc
gs + mi
12
2
pc+ep
pc + mi
11
ep + sy
ep + ca
a
1
mi + sy
b
p
c
d
e
f

III.5
Na_2Cl_2-K_2Cl_2-$MgCl_2$-Na_2SO_4-K_2SO_4-$MgSO_4$-H_2O

bi: bischofite ($MgCl_2 \cdot 6H_2O$), bl: bloedite ($Na_2Mg(SO_4)_2 \cdot 4H_2O$)
ca: carnallite ($KMgCl_3 \cdot 6H_2O$), da: d'ansite ($Na_{21}MgCl_3(SO_4)_{10}$)
ep: epsomite ($MgSO_4 \cdot 7H_2O$), gs: glaserite ($K_3Na(SO_4)_2$)
ha: halite (NaCl), hx: hexahydrite ($MgSO_4 \cdot 6H_2O$)
ka: kainite ($KMgClSO_4 \cdot 11/4\ H_2O$), ks: kieserite ($MgSO_4 \cdot H_2O$)
lg: langbeinite ($K_2Mg_2(SO_4)_3$), le: leonite ($K_2Mg(SO_4)_2 \cdot 4H_2O$)
lw: loeweite ($Na_{12}Mg_7(SO_4)_{13} \cdot 15H_2O$), mi: mirabilite ($Na_2SO_4 \cdot 10H_2O$)
pc: picromerite ($K_2Mg(SO_4)_2 \cdot 6H_2O$), sy: sylvite (KCl)
th: thenardite (Na_2SO_4), vh: vanthoffite ($Na_6Mg(SO_4)_4$)

halite and

1–a: mi + gs + sy
1–b: mi + pc + sy
1–12: mi + gs + pc
2–d: ca + ep + sy
2–4: ca + ep + ka
2–5: ep + ka + sy
3–c: ep + pc + sy
3–5: ep + le + sy
3–15: ep + le + pc
4–e: ca + ep + ks
4–9: ca + ka + ks
4–14: ep + ka + ks
5–17: ep + ka + le
9–h: ca + ks + sy
9–10: ka + ks + sy
10–i: ks + lg + sy
10–28: ka + ks + lg
11–p: ep + mi + pc
11–12: bl + mi + pc
11–15: bl + ep + pc
12–13: bl + mi + gs
12–16: bl + gs + pc
13–n: gs + mi + th
13–21: bl + gs + th
14–18: ep + hx + ka
14–19: hx + ka + ks
15–16: bl + le + pc
15–17: bl + ep + le
16–25: bl + gs + le
17–18: bl + ep + ka
17–22: bl + ka + le
18–20: bl + hx + ka
19–20: hx + ka + lw
19–28: ka + ks + lw
20–22: bl + ka + lw
21–m: da + gs + th
21–24: bl + da + gs
22–23: ka + le + lw
22–25: bl + le + lw
23–27: le + lg + lw
23–28: ka + lg + lw
24–l: da + gs + vh
24–26: bl + gs + vh
25–26: bl + gs + lw
25–27: gs + le + lw
26–29: gs + lw + vh
27–29: gs + lg + lw
28–30: ks + lg + lw
29–k: gs + lg + vh
29–30: lg + lw + vh
f–g: bi + ca + ks

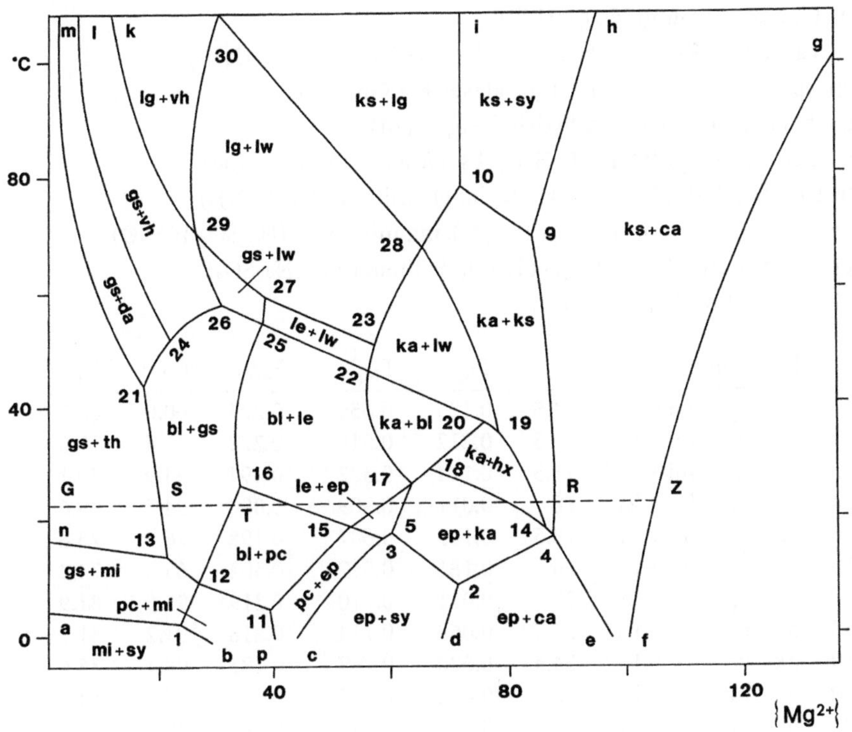

°C
80
40
0
40
80
120
{Mg2+}
m
l
k
i
h
g
30
lg + vh
ks + lg
ks + sy
lg + lw
10
80
gs+vh
29
9
ks + ca
gs + lw
28
27
gs+da
26
23
ka + ks
le + lw
25
ka + lw
24
22
21
bl + le
ka + bl
20
19
40
gs + th
bl + gs
ka+hx
18
16
17
G
S
le + ep
R
Z
T
n
15
5
ep + ka
14
13
3
4
bl + pc
gs + mi
12
pc+ep
2
pc + mi
11
ep + sy
ep + ca
a
1
d
e
f
mi + sy
b
p
c

III.5
Na_2Cl_2-K_2Cl_2-$MgCl_2$-Na_2SO_4-K_2SO_4-$MgSO_4$-H_2O

bl: bloedite ($Na_2Mg(SO_4)_2{\cdot}4H_2O$)
ca: carnallite ($KMgCl_3{\cdot}6H_2O$), da: d'ansite ($Na_{21}MgCl_3(SO_4)_{10}$)
ep: epsomite ($MgSO_4{\cdot}7H_2O$), gs: glaserite ($K_3Na(SO_4)_2$)
ha: halite (NaCl), hx: hexahydrite ($MgSO_4{\cdot}6H_2O$)
ka: kainite ($KMgClSO_4{\cdot}11/4\ H_2O$), ks: kieserite ($MgSO_4{\cdot}H_2O$)
le: leonite ($K_2Mg(SO_4)_2{\cdot}4H_2O$), lg: langbeinite ($K_2Mg_2(SO_4)_3$)
lw: loeweite ($Na_{12}Mg_7(SO_4)_{13}{\cdot}15H_2O$), mi: mirabilite ($Na_2SO_4{\cdot}10H_2O$)
pc: picromerite ($K_2Mg(SO_4)_2{\cdot}6H_2O$), th: thenardite ($Na_2SO_4$)
vh: vanthoffite ($Na_6Mg(SO_4)_4$)

	halite and	°C	F_{K_2}	F_{Mg}	F_{SO_4}	C_P	C_T
11:	bl + ep + mi + pc	4.5	0.125	0.650	0.225	60.0	68.5
12:	bl + gs + mi + pc	9.5	0.212	0.516	0.272	51.4	70.0
13:	bl + gs + mi + th	13.5	0.124	0.382	0.494	54.9	76.1
14:	ep + hx + ka + ks	18	0.011	0.909	0.080	94.7	89.6
15:	bl + ep + le + pc	18.5	0.115	0.687	0.198	76.4	75.8
16:	bl + gs + le + pc	26	0.181	0.517	0.302	65.3	74.4
17:	bl + ep + ka + le	27	0.071	0.710	0.219	91.0	81.9
18:	bl + ep + hx + ka	27.5	0.068	0.714	0.218	91.2	81.7
19:	hx + ka + ks + lw	34.5	0.020	0.797	0.183	98.1	85.5
20:	bl + hx + ka + lw	37	0.024	0.774	0.202	98.4	84.8
21:	bl + da + gs + th	44	0.260	0.345	0.395	49.3	74.0
22:	bl + ka + le + lw	47	0.112	0.658	0.230	84.5	80.9
23:	ka + le + lg + lw	52	0.114	0.663	0.223	85.3	81.9
24:	bl + da + gs + vh	52	0.257	0.385	0.358	55.6	76.5
25:	bl + gs + le + lw	55.5	0.231	0.492	0.277	75.8	82.1
26:	bl + gs + lw + vh	58.5	0.279	0.426	0.295	71.3	81.6
27:	gs + le + lg + lw	59	0.270	0.471	0.259	80.3	86.3
28:	ka + ks + lg + lw	68	0.131	0.719	0.150	90.1	88.4
29:	gs + lg + lw + vh	70	0.334	0.407	0.259	62.9	81.6
30:	ks + lg + lw + vh	108	0.262	0.499	0.239	56.5	80.8

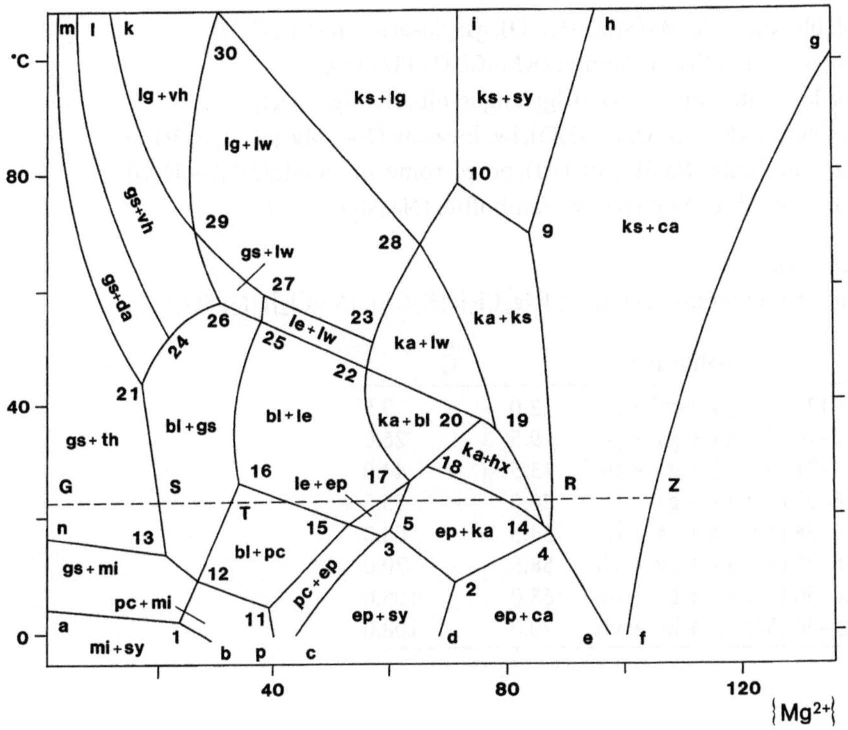
°C
80
40
0
40
80
120
{Mg2+}
m
l
k
i
h
g
30
lg+vh
ks+lg
ks+sy
lg+lw
10
gs+vh
29
9
ks+ca
gs+lw
28
27
gs+da
26
23
ka+ks
le+lw
24
25
ka+lw
22
21
40
bl+gs
bl+le
ka+bl
20
19
gs+th
ka+hx
18
16
G
S
le+ep
17
R
Z
T
n
13
15
5
ep+ka
14
3
bl+pc
4
gs+mi
12
2
pc+ep
pc+mi
11
ep+sy
ep+ca
a
1
d
e
f
mi+sy
b
p
c

III.5
Na_2Cl_2-K_2Cl_2-$MgCl_2$-Na_2SO_4-K_2SO_4-$MgSO_4$-H_2O

bl: bloedite ($Na_2Mg(SO_4)_2{\cdot}4H_2O$), gs: glaserite ($K_3Na(SO_4)_2$)
ha: halite (NaCl), ka: kainite ($KMgClSO_4{\cdot}11/4\ H_2O$)
ks: kieserite ($MgSO_4{\cdot}H_2O$), lg: langbeinite ($K_2Mg_2(SO_4)_3$)
le: leonite ($K_2Mg(SO_4)_2{\cdot}4H_2O$), lw: loeweite ($Na_{12}Mg_7(SO_4)_{13}{\cdot}15H_2O$)
mi: mirabilite ($Na_2SO_4{\cdot}10H_2O$), pc: picromerite ($K_2Mg(SO_4)_2{\cdot}6H_2O$)
th: thenardite (Na_2SO_4), vh: vanthoffite ($Na_6Mg(SO_4)_4$)

SALSYS
Input: temperature. Output: {Na_2Cl_2}, {K_2Cl_2}, {$MgCl_2$}, {$MgSO_4$}

	halite and		°C	
1–12 :	gs + mi + pc	2.0	-	9.5
12–16 (T):	bl + gs + pc	9.5	-	26.0
13–21 (S):	bl + gs + th	13.5	-	44.0
16–25 (T):	bl + gs + le	26.0	-	55.5
19–28 (Y):	ka + ks + lw	34.5	-	68.0
26–29 (V):	gs + lw + vh	58.5	-	70.0
28–30 (Y):	ks + lg + lw	68.0	-	108.0
29–30 (W):	lg + lw + vh	70.0	-	108.0

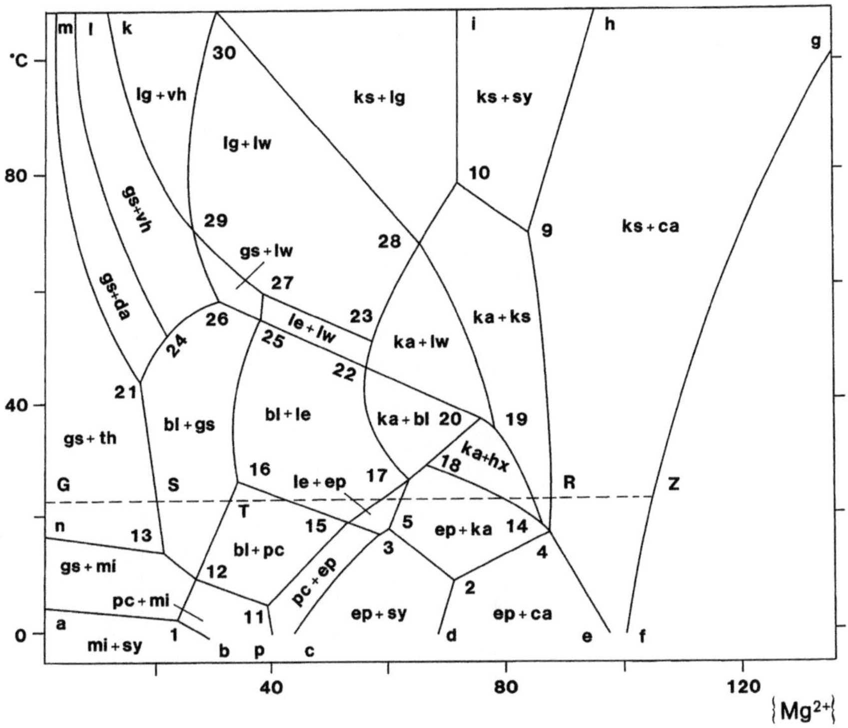

°C
80
40
0
40
80
120
{Mg2+}
m
l
k
30
i
h
g
lg + vh
ks + lg
ks + sy
lg + lw
10
gs+vh
29
9
ks + ca
gs + lw
28
27
gs+da
26
23
ka + ks
le + lw
24
25
ka + lw
22
21
bl + gs
bl + le
ka + bl
20
19
gs + th
ka+hx
18
16
le + ep
17
G
S
R
Z
T
15
5
ep + ka
14
n
13
3
bl + pc
4
gs + mi
12
pc+ep
2
pc + mi
11
ep + sy
ep + ca
a
1
mi + sy
b
p
c
d
e
f

IV Subject Index

MIX
Papier aus verantwortungsvollen Quellen
Paper from responsible sources
FSC® C105338

If you have any concerns about our products,
you can contact us on
ProductSafety@springernature.com

In case Publisher is established outside the EU,
the EU authorized representative is:
Springer Nature Customer Service Center GmbH
Europaplatz 3, 69115 Heidelberg, Germany

Printed by Libri Plureos GmbH
in Hamburg, Germany